Understanding Computer Telephony

Second Edition

How to Voice Enable Databases - From PCs to LANs to Mainframes

- Analog and Digital Phone Connections
- Caller and Called Party Identification
- Touchtone and Dial Pulse Detection
- Automatic Speech Recognition
- Digitized and Synthesized Speech
- Automated Fax Processing
- Data Access
- Computer Telephone Integration (CTI)

Carlton Carden

Published by Flatiron Publishing, Inc.
Copyright © 1997 by Carlton Carden

ISBN 1-57820-000-8

Manufactured in the United States of America

Second Edition, January 1997
Printed at BookCrafters, Chelsea, MI

Acknowledgments|

This book was made possible by the direct contributions of numerous technology partners.

Major thanks to Martin Zary of Dialogic Corporation who provided material for the chapters on Connecting to Analog Phone Lines, Connecting to T-1 Circuits, Caller and Called Party Identification, Touchtone/DTMF Detection and Serving the Hearing Impaired. Steve Mariconda also of Dialogic provided material for the chapter on Market Opportunities. Dan Zumar of Dianatel Corporation contributed to Connecting to ISDN Primary Rate Interface. Rafi Katz of Aerotel Ltd. in Israel contributed to Dial Pulse Detection.

Pete Foster of Voice Control Systems provided material for Automatic Speech Recognition. Andrew Machalik of Voice Information Systems, Inc. provided material for Digitized Speech. Sara O'Malley and Elizabeth Peters of Berkeley Speech Technologies, Inc. contributed to Speech Synthesis. Maury Kauffman of The Kauffman Group and Rosie Pongracz of GammaLink provided material for Automated Fax Processing.

Pru Heikkinen and Lynwood Taylor of Interface Systems, Inc. contributed to Accessing IBM Mainframe Data. Mark Proudfoot of Micro-Integration Corp. provided material for Accessing IBM System/3X and AS/400 Data. Accessing

SQL Data with ODBC was drawn from the Microsoft Backgrounder Accessing the World of Information - Open Database Connectivity (ODBC), Microsoft's Software Developer Kit (SDK): ODBC 2.0 Programmers Reference and INTERSOLV's DataDirect ODBC Drivers Reference and Benchmark. Carl Strathmeyer of Dialogic contributed material for Computer Telephone Integration. Versit was drawn from the Computer Telephony Integration White Paper (versit CTI Encyclopedia.

Patrick Gilbert of Expert Systems, Inc. provided the sample order status application and the EASE coding examples. Alan Hansen, Joel Heysel, Eric Lyons, Mark Nacon, Matt Stuit, and Jerry Wilson also of Expert Systems contributed to the review and editing process. Joy and Spike Carden, friends, family, and all the people at Expert Systems provided patience and support in this time consuming effort.

And last but not least, Harry Newton of Computer Telephony Magazine contributed definitions for the Glossary from Newton's Telecom Dictionary.

Many thanks to the people who provided material. Without these contributions, this book would not have been possible. Of course, errors and omissions are the sole responsibility of the author.

Company and product names used in this book are trademarks or registered trademarks of their respective organizations.

The author is president of Expert Systems, Inc. - maker of EASE computer telephony software and EASEy FrameWorks application templates. Expert Systems and the author are therefore active participants in the computer telephony industry and cannot claim to be impartial.

Expert Systems is an Open Toolkit Developer and CT Select partner with Dialogic Corporation. Expert Systems is also a Microsoft Solution Provider and an INTERSOLV Channel Partner. Expert Systems and the author work closely with technology partners who contributed material for this book.

While every effort has been made to ensure accurate information in this book, neither the author nor the publisher may be held liable for errors in printing, or inaccurate or incomplete information. This book is sold "as is" without warranty of any kind, expressed or implied, of merchantability or fitness for a particular purpose.

The reader is encouraged to verify all relevant information with vendors when making purchase, design or business decisions.

The author and publisher welcome suggestions and corrections. You may contact the author or publisher at the following addresses:

Carlton Carden
Expert Systems, Inc.
1301 Hightower Trail
Suite 201
Atlanta, GA 30350
Tel: 770.642.7575 x125
Fax: 770.587.5547
Email: info@EASEy.com

Christine Kern
Flatiron Publishing, Inc.
12 West 21st Street
7th Floor
New York, NY 10010
Tel: 212.691.8215
Fax: 212.691.1191

What wonderful progress would humanity make if foolish and false ideas in philosophy of life and political economy eliminated themselves, with their creators, as completely as the mistakes of a pilot-designer eliminated him and his machine.

-Igor Sikorsky, Aviation Pioneer

Table of Contents|

Foreword|

Harry Newton
Editor-in-Chief
Computer Telephony Magazine

The Surround Theory – Giving Data A Voice

Think about your overnight package. What do you care about? You care it got there. Let's say, it didn't. You go ballistic. You call the package delivery company. They say they'll check and "get back to you." Click. Not what you want to hear. You want to hear where the package is and when it will be delivered.

You see an ad for a great pair of shoes. You want a pair. Where to buy them?

You took your car in for service. They promised it would be ready by Wednesday afternoon. Is it?

You eye a gorgeous coat in a shop window. Big sale. Big deal for cash. Do you have enough money in your checking account?

Products and services are largely the same, these days. Packages get reliably delivered by many overnight deliverers. What makes the BIG difference is something I call "The Surround Stuff." All the information that "surrounds" the product. Increasingly, people want to pick up the phone and dial for information.

It would be great if the world could afford 24-hour a day slaves answering boring questions, like "When will the package be delivered?" "The closest shoe store to you is on West 23 Street." "You have gobs of money in the bank."

But the world can't. Moreover, people don't like sharing their intimacies (their bank balances) with clerks who snicker. People don't want to waste time on the niceties of social stuff. Just gimme the info, fast.

Increasingly, "the info" sits in a computer somewhere. Increasingly, people are calling that computer with their phones and asking it for simple answers. They do their "asking" by punching touchtone buttons. Or they "speak" to the computer, which listens with new technology called automatic speech recognition.

Carlton Carden, who wrote this book, calls this new business "Giving Data a Voice." I wanted to call this book "Giving Data a Voice," but Carlton said "No way. No one will know what we're talking about."

But the book is about giving data a voice. It's an area Carlton and his firm Expert Systems in Atlanta know a lot about. They wrote some of the seminal software in the area. They've probably given more databases a voice than any other single entity in the western world (with few exceptions).

Read this book. You'll feel comfortable about voice enabling your databases – whether they're located in single PCs, or on servers attached to LANs (local area networks), or in mainframes.

The book covers everything from how to get to the information in your databases, to how to present it to your customers in forms most useful to your customers – from the spoken word to a fax.

Study the material in this book. Follow its advice. I promise that your customers will love you far more than they do now.

Enjoy and learn.

Harry Newton
New York
January, 1997

I|
Opportunities and Applications

1|
Introduction

2|
Market Opportunities

3|
Sample Application

1|
Introduction

Premise of this Book

The purpose of this book is to help you understand how *Computer Telephony* works. By the way, telephony is pronounced 'teh-**lef**-uh-nee', and italicized terms are defined in the Glossary. If you see an acronym you don't recognize, it is probably decoded in the Glossary as well.

As a teaching aid, I will outline an example of 'Giving Data A Voice' - an order status inquiry application. This book will examine the various technologies that can be used to implement the sample application.

The examination of Computer Telephony components is organized into five sections:

- Connecting to the Telephone Network

- Caller Input

- Computer Telephony System Output

- Accessing Data

- Computer Telephone Integration

In each chapter, I present the fundamentals of the technology so that you understand WHAT IT IS and HOW IT WORKS. Then in most cases, I illustrate how the technology can be implemented in our sample order status application.

First, let's take a look at the market opportunities in Computer Telephony.

Market Opportunities

Welcome to Computer Telephony (CT)

Virtually everyone has read about computer telephony, or experienced it first hand by checking their bank balance using a touchtone telephone (figure 2-1).

Typical Computer Telephony Application

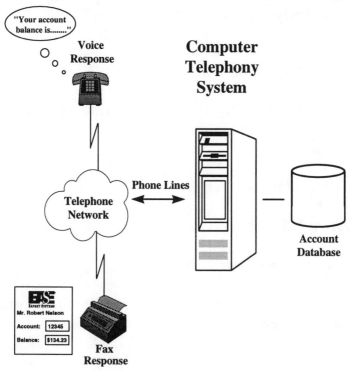

Figure 2-1

But what exactly is computer telephony?

Computer telephony is the general term used to encompass a wide variety of technologies and applications that use the power of a computer (often a server on a local area network) to add intelligence to telephony functions, and mesh these functions with data processing.

Today, telephony functions such as receiving and sending messages, routing telephone calls, and performing actions based on touchtones entered by a caller can be controlled by a voice processing server. This overlay of computing intelligence on telephony functions makes possible applications that streamline many common business practices. Customer service is improved. Sales people are more productive. Employees are more efficient.

Intended Audience

Since you are reading this book, I assume you or your organization fit into one or more of the following categories:

- Value Added Reseller (VAR)

- System Integrator

- Software Developer

I also assume that you are investigating opportunities in the computer telephony market.

Market Research Highlights Opportunities

CMP Publications, Inc., publisher of VARBusiness and Computer Reseller News, conducted research on the end user market for computer telephony systems in October of 1995.

CMP interviewed 300 MIS decision makers from organizations with between ten and 999 employees. Each of the organizations purchased products and/or services from VARs or system integrators. The average organization had 20 sales people and 31 customer support representatives. The leading vertical markets were Manufacturing (44%), Health Care (10%), Wholesale/Distribution (4%), Financial (4%) and Retail (4%). 77% had local area networks (LANs) installed.

Key survey results were:

• 9% have already invested in computer telephony systems.

• 58% will have invested within the next two years.

The top three computer telephony benefits and the percentage of respondents who considered each to be "Very Important" to their organization:

• 65% - Respond to caller inquiries with consistent and accurate information.

• 56% - Gain a competitive edge.

7

- 48% - Increase staff productivity by automating repetitive functions.

Open Standards Computer Telephony

Until a few years ago the majority of computer telephony systems were "closed": they were proprietary, "black box" turnkey solutions sold only by phone companies, proprietary voice processing manufacturers, and telephone interconnect companies.

Today, most CT systems are built on "open" architectures. Similar to the way multiple suppliers can create hardware and software for IBM compatible PCs, the open architectures have allowed for a number of published standards for building PC-based CT systems.

These standards let you mix and match LAN interface cards, voice/fax processing and telephone interface boards from companies such as Dialogic Corporation, and CT application software such as EASE from Expert Systems, Inc. that interacts with other applications and databases. And because the CT components are standards-based, you can be sure they will all work together.

Open CT standards have fostered today's synergistic relationship between telecommunication and data processing technologies. Voice communication is no longer isolated from information systems, so both technologies are more useful, more powerful, and more accessible than ever before. The speed and power of the data processing leverages the convenience and omnipresence of the worldwide telephone network.

Why Should You Add Computer Telephony to Your Portfolio?

As a system integrator or VAR, there is an array of technologies competing for your attention—the Internet, document imaging, wireless LANs, and so on. As in any business, you have limited technical and financial resources, so you must invest wisely. What makes CT stand out as a technology and as a business opportunity? Simply this: all your customers have telephones, and they all have customers of their own.

The telephone is, of course, the most widely used business tool. Go in any office, factory, distribution center, hotel, restaurant, bank, retail store, hospital, or home, anywhere business is conducted, and you are likely to find a phone in most rooms. Computer telephony turns all these phones into virtual "terminals". The telephone keypad becomes a keyboard that callers can use to send and receive information and instructions. It saves callers time, and saves the called party - your customers money.

A large proportion of employee time is spent on the telephone. In particular, all your customers have customers of their own to deal with. It is this fact that makes computer telephony such a compelling opportunity. It enhances existing business activities and applications in a way no other technology can.

All businesses, regardless of industry, must do certain things: communicate with prospects, provide information about products and services, take orders and answer customer inquiries. All your customers want to make their employees--in particular, their sales people and customer service people--more productive. CT does this across the board. You can add value by customizing CT software solutions to address unique vertical market or situation-specific buyer requirements.

CT Makes it Easy for Your Customers to Say "Yes"

Because CT can streamline and enhance a wide variety of common business processes -- communicating with employees and customers; improving the productivity of sales people; and improving customer service--it is easy to sell. You can approach buyers from a needs assessment perspective, and make it easy for them to say yes to the CT solution you propose.

Here are some examples of questions you might pose to a CT prospect:

- **Do You Want to Improve Customer Service?** Most of us know the benefits of banking by phone, and perhaps the attendant sense of relief from not having to stand in line and deal with a teller during inconvenient "banker's hours." Now your clients can provide their customers similar benefits. Some clients, for example, may wish to allow callers to check inventory availability by phone, make a credit card payment, or find out the status of their order. All this can be done by the caller during a single call, with no agent assistance. When callers place frequent and predictable orders they may use the telephone to do so. By adding computing power to telephone functions, computer telephony also automates getting information to callers. Callers can, for example, be given access to an audio bulletin board that provides round-the-clock information about product availability, or answers to frequently asked questions. They can also receive directions to an office or branch, or find the dealer nearest them based on their zip code. These transactions can be confirmed with interactive fax.

- **Do You Need to Improve Communications with Your Employees?** By adding computing power to telephone functions, your customers can streamline internal communications with their employees. For example, home health care personnel can "punch in" and "punch out" of work using the telephone keypad, and the system can capture the telephone number from which they called. Your customer can enable employees to dial into a cafeteria style benefits system that provides recorded information about benefit options, and allows the employee to modify their benefit program choices using their telephone keypad. This supports the use of human resources in the most efficient way possible and improves employee satisfaction.

- **Do You Need to Enhance the Efficiency of Your Employees Who Deal With Customers?** Computer telephony can also enhance the efficiency of those employees like sales people and customer support representatives, who interact with customers by phone. Everyone you sell to has customers of their own -- and can benefit both financially and operationally from streamlining the way employees interact with customers. CT systems can gather information from a caller on hold -- such as their account number. Based on this input, the caller can be identified and routed to the agent that is best qualified to handle the call. In addition, callers can assemble their own literature pack by selecting documents from a fax-on-demand system.

LAN And Database Interaction

More and more of this billion dollar market consists of smaller systems--particularly for providing callers with touchtone database access. VARs and system integrators are expected to become a dominant distribution channel because of strong customer relationships regarding their networks and databases.

The Vertical Market Perspective

Though there is a fair amount of overlap in the reasons businesses or industries would use CT, each could also have its own specialized uses. That's part of the value of a CT system; it can be easily customized to address your clients' specific needs, whether they are order entry, account status checking, routine data entry or inquiry, or any of the technology's myriad uses. Vertical market software developers may consider CT as an Input/Output peripheral that should be offered as an extension to their current product line.

Why Should You Add Computer Telephony to Your Portfolio?

Here are a few of the compelling reasons:

- **Your Customers Want It.** As mentioned above, it is the ubiquity of the telephone and the widespread operations (like customer service) all your customers wish to streamline that makes CT valuable. To date, there has been no one who can fulfill the sales need. Perhaps you've already been asked about it. Today, you're in a position to address these requirements, and enhance your margins in the process. Even as you read this, you have no doubt thought of several clients that you could call right now--clients seeking solutions to solve the business problems that CT addresses. A few moments reflection, or a conversation with a sales person, will bring more near-term opportunities to mind. A coordinated sales effort to your customer base is an easy way to generate a revenue spike with the business that will result.

- **An Extension of What You Already Sell.** If you sell database applications, CT lets you provide remote access to data using touchtone entry. Order entry systems can be enhanced to let outside sales people or even customers enter orders by phone. The possibilities to add value to new or existing customer service applications, help desk applications, human resource systems, inventory management systems, or any application--is here now.

- **Foster a "Business Reengineering" Posture with Key Customers.** Computer telephony can help position you as a consultant who can make your clients' businesses run better in many ways. It increases your value as a business partner, and gets you "inside" your clients' operations in a way that fosters long-term relationships.

- **Open, Standards-Based Solutions.** There are more options available to developers and resellers of CT solutions than ever before. Companies like Dialogic Corporation, along with their business and technology partners, have pioneered the interoperability that makes component CT a reality today. This means you can offer your clients field-proven solutions that are still state-of-the-art.

- **Easy to Cost Justify.** CT is easy to cost justify to your clients, because there are hard benefits (like savings in payroll costs when customer service head count growth is controlled) as well as soft benefits (like improved customer satisfaction and employee productivity). Growing personnel costs and other labor issues, as well as a desire to serve as many customers as possible as efficiently as possible, mean that CT will continue to grow in importance as a key enabling technology for business in the years ahead. And CT costs continue to decline, so even small companies are able to justify CT systems and solutions. It is easy to show clients how CT provides value for their business, helps them become more efficient, and saves them money.

- **Demand High, Competition Low.** Offering CT today, while the demand is high but the number of qualified suppliers is limited, can allow you to secure higher margins. CT is an industry that everyone agrees will continue to grow explosively, but because it is new to the VAR channel, there is a window of opportunity to establish yourself as a leader in your geographical or vertical market. Being the first expert on CT solutions in a given market will provide a competitive advantage you can leverage for years to come.

- **The Applications are Here.** High level, Windows and Windows NT based application generators such as EASE and application templates such as EASEy FrameWorks from Expert Systems, Inc. mean an abundance of CT applications that are easy to configure, install, modify, support, and enhance. Whether the customer requirement is account inquiry, fax-on-demand, specialized voice messaging, or a totally custom application, current CT software tools provide pain relief and profits.

Your Business and Financial Rationale for CT

Once you accept the premise that computer telephony integration is a technology whose time has come, and one for which the market is vast and largely untapped, you are likely to ask these next questions:

- "How quickly should I act?"

- "How do I get started?"

The answer to the first question is "Immediately." The rationale for acting swiftly is convincing.

15

The answer to the second question is hopefully contained in the Appendix of this book where the name, address, phone and fax numbers for each software and hardware component supplier profiled are listed.

3

Sample Application

Demonstration Line

Expert Systems, Inc. has provided a sample interactive voice/fax response application that will be used throughout this book to illustrate the various technologies that comprise computer telephony. For a demonstration of the order status application, follow these directions:

- Dial 770-642-7575

- A voice mail/automated attendant system will answer

- Enter extension 300 for the demonstration system

- When the system answers, follow the instructions

- Press 1 to check the status of an order

- Enter any 5 digit order number (e.g. 12345)

- You will hear the status of your fictitious order

- Continue to follow the instructions, and you can request a fax confirmation of the fictitious order

Note: The actual vocabulary may differ slightly from the call flow in figure 3-1 and outline to follow.

Order Status Application Flow Chart

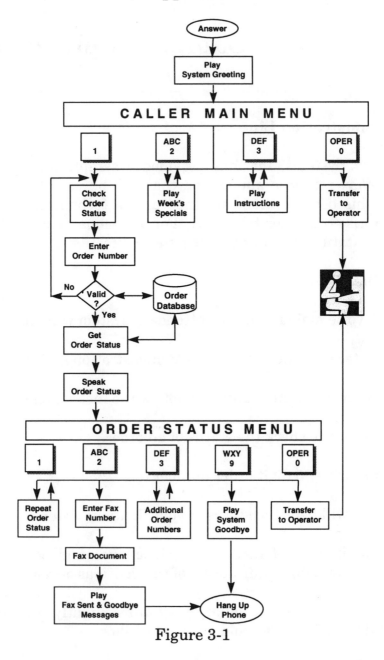

Figure 3-1

Order Status Application Call Flow

The following outlines the major nodes of call flow logic for our order status application:

ANSWER
When ring is detected, take line off hook.
Go to PLAY GREETING MESSAGE

PLAY GREETING MESSAGE
"Thank you for calling the ABC Company."
Go to CALLER MAIN MENU.

CALLER MAIN MENU
"To check the status of an order, press 1.
   For information on this week's specials, press 2.
   For order status instructions, press 3.
   To speak to a customer service representative, press 0."

If 1, go to ENTER ORDER NUMBER.
If 2, go to PLAY WEEK'S SPECIALS.
If 3, go to PLAY INSTRUCTIONS.
if 0, go to TRANSFER TO OPERATOR.

ENTER ORDER NUMBER
"Please enter your 5 digit order number now."
Get 5 digits, store in variable Order Number, then go to GET ORDER STATUS.

PLAY WEEK'S SPECIALS

Play weekly special message, then go to CALLER MAIN MENU.

PLAY INSTRUCTIONS

"To check the status of your order, you will need to enter an Order Number. For your convenience in this demonstration, you may enter any five digits as an Order Number."

Go to CALLER MAIN MENU.

TRANSFER TO OPERATOR

"Please hold while I transfer you to the next available representative."

Transfer to CSR hunt group using intelligent call progress analysis to assure completion. Go to HANG UP PHONE.

GET ORDER STATUS

Submit variable Order Number to database source and retrieve variables for Order Status, Order Amount and Ship Date.

If Order Status equals Shipped, go to SHIPPED STATUS.
If Order Status equals Scheduled, go to SCHEDULED STATUS.
If Order Status equals Canceled, go to CANCELED STATUS.
If Order Status equals Hold, go to TRANSFER TO ACCOUNTING

SHIPPED STATUS
"Your order number..."
speak Order Number as individual digits...
"in the amount of..."
speak Order Amount as dollars and cents...
"was shipped on..."
speak Ship Date as month, day, year.
Go to ORDER STATUS MENU.

SCHEDULED STATUS
"Your order number..."
speak Order Number as individual digits...
"in the amount of..."
speak Order Amount as dollars and cents...
"is scheduled to ship on..."
speak Ship Date as month, day, year.
Go to ORDER STATUS MENU.

CANCELED STATUS
"Your order number..."
speak Order Number as individual digits...
"in the amount of..."
speak Order Amount as dollars and cents...
"has been marked as canceled."
Go to ORDER STATUS MENU.

TRANSFER TO ACCOUNTING
"We have experienced a problem processing your order. Please hold while I transfer you to the next available representative."
Put caller on hold. Dial accounting hunt group using intelligent call progress analysis to assure completion.

When an accounting representative answers, the system will say: "Hold status transfer on order number..." speak Order Number as individual digits.
Complete call transfer. Go to HANG UP PHONE.

ORDER STATUS MENU
"To repeat this information, press 1.
For fax confirmation, press 2.
To check another order, press 3.
To end this call, press 9.
For a customer service representative, press 0."

If 1, go to GET ORDER STATUS.
If 2, go to ENTER FAX NUMBER.
If 3, go to ENTER ORDER NUMBER.
If 9, go to GOODBYE.
If 0, go to TRANSFER TO OPERATOR.

ENTER FAX NUMBER
"Please enter your area code and fax number."
Get 10 digits and store in variable Fax Number.

If Order Status equals Shipped, go to FAX SHIPPED STATUS.
If Order Status equals Scheduled, go to FAX SCHEDULED STATUS.
If Order Status equals Canceled, go to FAX CANCELED STATUS.

FAX SHIPPED STATUS
Merge Order Number, Order Amount and Ship Date with Fax Template: Shipped.

Send fax to Fax Number.
Go to FAX SENT MESSAGE.

FAX SCHEDULED STATUS
Merge Order Number, Order Amount and Ship Date with
Fax Template: Scheduled.
Send fax to Fax Number.
Go to FAX SENT MESSAGE.

FAX CANCELED STATUS
Merge Order Number and Order Amount with Fax
Template: Canceled.
Send fax to Fax Number.
Go to FAX SENT MESSAGE.

FAX SENT MESSAGE
"Your fax confirmation will be sent soon. If your fax line is
busy, three more attempts will be made. If we are unable
to get through after four attempts, you will need to
request another fax confirmation."
Go to GOODBYE.

GOODBYE
"Thank you for calling. Goodbye."
Go to HANG UP PHONE.

HANG UP PHONE
Hang up phone, then go to ANSWER and wait for next
call.

Other Applications

Although the Order Status application will be used throughout this book to educate you on how the components of Computer Telephony work, the universe of possible applications is limited only by your imagination.

Here's a sample list of typical applications:

- Order Entry

- Inventory Availability

- Account Inquiry

- Claim Status

- Dealer Locator

- Product Locator

- Shipment Tracing

- Warranty Registration

- Benefits Enrollment

- Product and Service Information

- Employee Scheduling

- Student Registration

- Rate Quoter

- Loan Calculator

- Talking Classified Ads

- Job Postings

- Homework Hotline

- Event Schedules

- Remote Time Clock

- Remote Payroll Entry

- Music Sampler

- Voice Dialing

- Automated Paging

- Integrated Voice Messaging

- International Callback Service

- Prepaid Calling Card Service

For More Information on Applications

For more information on typical applications, refer to <u>236 Killer Voice Processing Applications</u> by Edwin Margulies published by Flatiron Publishing, Inc.

II|

Connecting to the Telephone Network

4|
Connecting to Analog Phone Lines

5|
Connecting to T-1 Circuits

6|
Connecting to ISDN Primary Rate Interface

7|
Connecting to E-1/Euro ISDN Circuits

4

Connecting to Analog Phone Lines

Why is Telecommunication Terminology so Arcane?

If you've ever been in the position of ordering telecommunications services for your business, you're probably aware of the confusing array of connections available - Loop Start, Ground Start, T-1, ISDN, etc.

The confusion is exacerbated by the arcane terminology of the telecommunications world. You must realize that the US telecommunications industry existed in total isolation from competition and other industries up until *divestiture* in 1983, when AT&T was forced to spin off the seven *Regional Bell Operating Companies* (RBOCs). In this pre-divestiture void, the telecommunications industry developed a language that was completely foreign to the rest of the world. Its not unlike discovering a completely new country. Even a formally trained electrical engineer from the computer industry has difficulty conversing with their counterpart from the telecommunications industry.

Connecting to the Telephone Network

I just plug my CT voice processing board into the local wall jack, right? Yes, in some cases it is that simple, but then again it's not. Remember, a computer telephony system puts the power in your hands. Therefore, there are some things you need to know. This section serves a few purposes:

- Introduce the necessary telecommunication terminology.

- Explain the basic operation that you will be controlling.

- Help you select the most appropriate telephone service from the array of offerings.

One consideration is whether to connect the CT system directly to the telephone network, or to the local business telephone system (PBX or Key system). For smaller systems, 8 lines and below, it is probably more cost effective and more powerful to connect through a business telephone system, if you have one. In either event, you will need to understand the operation of the basic analog loop start telephone line.

Connecting Directly to the Telephone Network

So What's Behind the Wall Jack?

Loop Start is the telecommunications term for the type of telephone line which you are most familiar with; the type that usually services your home or small business, and works with standard telephones that you can buy almost anywhere. There are many different types of telephone lines, Ground Start, T-1, ISDN, etc.; but, loop start lines are the most technologically simple and the most ubiquitous throughout the world.

A loop start line consists of a single pair of wires between your telephone and the local telephone company's *central office* (CO). Loop start is also referred to as an *analog phone connection*, as opposed to T-1, E-1 and ISDN, which are "digital" services. Analog loop start lines are also often referred to as "the local loop" because this type of circuit is not used within the telephone network, only between the local CO and the customer.

Wall Jack and RJ-11 Modular Plug

RJ-11-4

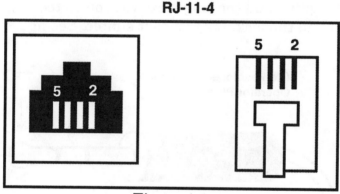

Figure 4-1

As illustrated in figure 4-1, if you examine your wall jack, and the *modular plug* (RJ-11) which connects to it, you'll notice four wires. The red and green wires are the telephone line to the CO. The black and yellow wires were originally used to supply power to light the dial on Trimline phones

Today, many voice boards use these "outer pair" of wires (the black and yellow) to carry a second telephone line out of the board. These jacks, which are physically identical to the RJ-11, are called *RJ-14*. You'll need a converter module from a telephone supply store (such as Radio Shack) which will connect the black/yellow to a second pair of red/green (RJ-14 to two RJ-11 converter).

Each of the wires in the loop start pair actually have a name - *tip* and *ring*. This is the telecom industry's way of saying plus and minus, or ground and positive in an electrical circuit.

The names date back to the days of the telephone operator's cord board. The tip wire was connected to the tip of the plug, and the ring wire was connected to the slip ring around the jack. It's just that simple. See figure 4-2.

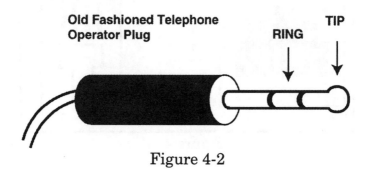

Old Fashioned Telephone Operator Plug

RING

TIP

Figure 4-2

Hopefully you are starting to see that telecommunications is not as complex as you first thought. The early high priests of telecom just made up mystical names to confuse the uninitiated.

Some Telecom Basics

To the telephone company, you are a *subscriber* to telephone service, and they supply you a *subscriber line*. The telephone company typically owns the wires up to your home or business, and you own the telephone wires and equipment within. Somewhere between your wall jack and the CO is at least one junction box. This is where your wiring joins the telephone company's wiring. This represents a *demarcation point* (see figure 4-3). When you order service, the telephone company provides the service up to the demarcation point, and you are responsible for connection from that point on.

Central Office, Local Loop, and Demarcation Point

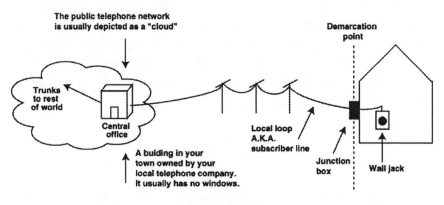

Figure 4-3

So what's to prevent a subscriber from connecting any old home brewed device up to the telephone network? Every country has a regulatory body that governs devices which connect to the *public switched telephone network* (PSTN).

Typically, it is your responsibility to ensure that every device that you connect to the public network is approved by the appropriate governing body, which in many countries is still the state run, regulated monopoly (many times referred to as the *Postal, Telegraph and Telephone* administration, or PTT).

In the de-regulated environment of the US, the FCC is the governing body, and telephone devices must have *Part 68 approval*. Look for a label on the voice board that indicates the approval numbers, and if you're outside the US make sure the vendor offers a board that is approved for use in your country. Some countries, such as Japan, require system level as well as board level approvals.

You may notice that there are a large bunch of wires coming from the CO into your junction box. Only two are used for each telephone line that you "subscribe" to. The others are for spares and for extra phone lines which you may wish to order. The local phone company isn't about to string another cable or dig another trench each time a resident orders another phone line.

Placing an Outbound Call

The term "loop start" gets its name from the signaling protocol which is used to initiate a call. The CO has a battery at their end of the line (they really don't use batteries too much any more, but instead an integrated circuit that electrically looks like one). The battery is typically 48 volts direct current (DC). Your end has a switch, called a *hookswitch* or *switchhook*, which is typically controlled by the plunger underneath the telephone handset.

Idle State

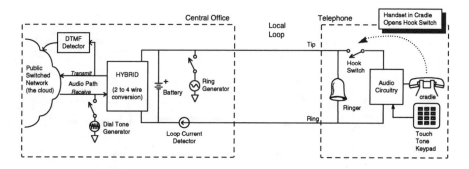

Figure 4-4

As illustrated in figure 4.4, when the telephone handset, or receiver, is on-hook, the plunger is depressed and the hookswitch is open, creating an open circuit or loop between you and the phone company. No current flows from their battery.

When the handset is lifted off-hook (figure 4-5) and the plunger pops up, the hookswitch is closed and a circuit loop is then created between the phone and the CO. Direct current flows and you have *loop current*. The CO detects the flow of current and if it's ready to accept your call, *dial tone* is presented.

Dial tone is a continuous tone produced by combining two different frequencies (350Hz and 440Hz in North America). Your lifting the handset is viewed as a request for service (also called a *seizure*), and the dial tone is a signal that you have been granted service. Therefore, you start the call by creating a loop... hence , "loop start".

Placing an Outbound Call

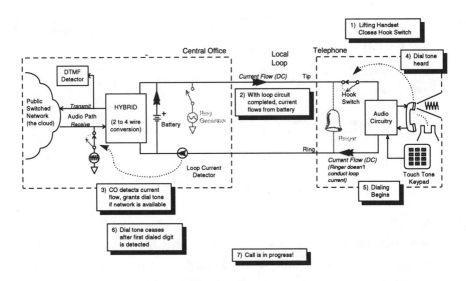

Figure 4-5

The CO then expects you to begin dialing. There are two types of dialing used on loop start lines. *Tone dialing* uses sound to represent the digits 0 through 9 and the characters # and *. Each digit is assigned a pair of frequencies (Dual Tone Multi-Frequency or DTMF, see Chapter 9). *Pulse dialing* is an older method used by rotary telephones that represents digits by interrupting the DC loop current (using the hookswitch) to produce "pulses." The number of pulses equals the digit dialed (see Chapter 11).

In the US, we take the dial tone for granted. We lift the handset and the dial tone is there. Not unlike the refrigerator light. However, if everyone in our local community lifted the handset at the same time, not everyone would get dial tone. Such is also the case in some countries where phone service availability has not kept pace with demand. So why is this important? Many voice boards don't look for dial tone before dialing out. They go off-hook and begin dialing after a pre-set period of time, assuming that dial tone is present - call this blind dialing.

This works OK in the US and many other markets. However, some countries have laws against blind dialing. Why? Suppose the number you intend to dial is 555-2911. The voice board begins dialing before the CO is ready. The CO suddenly becomes ready and presents dial tone just after the 555-2 is dialed. You have just dialed 911 - the US emergency service number. Explain that to the local authorities. This explains why you should set your voice board to wait for dial tone before dialing.

Voice boards go beyond just dialing out. They can monitor and report the result of an outbound call - whether it reached a busy signal, a ring-no-answer, an operator intercept, a fax machine, an answering machine, or a human voice answer. This is a simple task for humans but a challenge for computerized voice boards.

Hookflash Transfer

Why would an CT system need to make an outbound call? It happens in the event that a caller requests operator assistance. The telephone side can signal the CO in the middle of a call by momentarily going on-hook, then back off-hook again. This is called a hook-flash.

The CO detects the momentary loss of loop current, and places the caller on hold, and applies a dial tone to your phone. You can then place an outbound call to a new destination. Another hookflash signals the CO to transfer the call to the new destination. This feature is not standard on a loop start line, you'll need to request this service from the phone company (*Centrex* service). Most in-house phone systems (PBX, key systems) support this hookflash transfer feature.

The timing of a hookflash is important. If the duration is too long (usually greater than 1 second) the CO thinks that you ended the call by hanging up. If the duration is too short (usually less than 3/10ths of a second), the CO doesn't see the signal. The voice board's default setting will work in most cases. You'll find many new phones on the market today have "flash" buttons. This button merely performs the hookflash, but takes the human error out of the timing aspect.

In some parts of Europe, the mechanism for signaling the CO for transferring calls is called earth recall. Earth recall is electrically different than a hookflash, but accomplishes the same function. Any voice board will do a hook flash, but not all will do an earth recall. Check your needs carefully.

For more details on call transfer, see the Chapter 16 - Call Transfer and Outdialing.

Receiving Inbound Calls

As we've seen in the previous section, you signal to the phone company that you wish to make a call by loop signaling. When you have an incoming call the CO signals to you through ring signaling (figure 4-6).

Receiving a Call - Ring

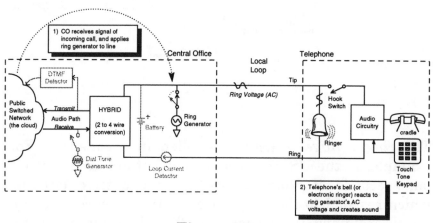

Figure 4-6

CO's have banks of ring generators, which provide 30 to 120 volts AC. To signal a phone, the ring generator's output is applied to the CO side of the line. The telephone has a bell ringer, more likely an electronic ringer on an modern phone, which is wired in parallel with the hookswitch.

When the telephone handset is lifted off-hook, the CO detects the current flow and ceases ringing. This is called *ring trip* (figure 4-7).

Receiving a Call - Answer

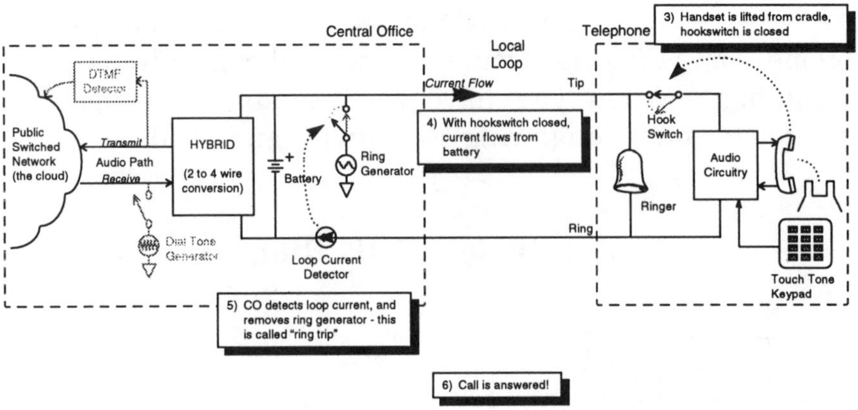

Figure 4-7

The *ring voltage* is usually applied in a *cadence*. For example, in the US, that cadence is a standard 2 seconds on, 4 seconds off. Voice boards have ring detector circuits, which respond to the ON portion of the cadence, and can be programmed to answer (go off-hook) after a desired number of rings. For a CT application, the setting should be as small as possible, to give the appearance to the customer that the call is being answered promptly.

Some voice boards must wait until the end of the "ring on" cycle to avoid exposing the delicate audio circuitry to the high power ring signal. Look for a voice board which allows you to answer the call anytime during the ring cycle.

The CO's ring generators are continuously following a cadence. The cadence does not start when a call is initiated. Instead, the generator is applied to the line without regard for the state of the cadence. This causes the first "ring on" cycle to appear to the phone to be anywhere between 0 and 2 seconds.

This system results in an occasional error condition. You may receive an incoming call to your line, and the ring generator is applied during the off cadence. It may be as much as 4 seconds before the phone rings. What if, during this 4 seconds you decide to place an outbound call and lift the handset? The incoming call is complete and you are left puzzled.

This error condition is called *glare*. Loop start lines are inherently susceptible to glare, and there is nothing that you or a voice board can do to prevent it. There exists a variation on the loop start protocol known as ground start. However, most voice boards do not support this, and you need to place a special order (and pay more) to your phone company to get a ground start line.

Also be aware that the incoming ring voltage is somewhat hazardous. Remove the telephone line from the voice board before handling it for installation or removal. Even if the power is off on the PC, a ring voltage coming in via the phone line can shock you.

How Do I Know the Call is Complete?

Telephone calls between polite people end in "good-bye." However, we have all experienced the uncomfortable situation of listening to a short period of silence before realizing the other party has hung up. Since CT systems don't usually understand good-bye, they rely on

41

disconnect notification, general tone detection, time-out or silence detection.

- **Disconnect Notification:** If the CO provides disconnect notification, the CT system can immediately hang up the line and prepare for the next call. Disconnect notification is a momentary reversal or removal of the loop current provided by the CO. This change in battery condition is detected by the voice board and passed to the CT application.

- **General Tone Detection:** Some countries provide Disconnect Supervision in the form of a continuous tone to indicate that a party has hung up. In Japan for example, you will receive a 400 Hz tone. A voice board with capabilities such as Dialogic's general tone detection can detect this signal as a hang up notification. General tone detection can also be used to look for new dial tone presentation which is often an indication of caller hang up.

- **Time-out:** If battery reversal or tone notification is not provided, the CT application can set time limits for periods of inactivity. If this time limit is exceeded, the CT application assumes that the other party has hung up.

- **Silence Detection:** In certain situations such as voice message recording, the voice board can be instructed to listen for a minimum interval of silence which it will consider as a hang up. In addition, the voice board can be instructed to periodically sample the phone line to detect a minimum interval of silence which will be considered a hang up.

Connecting Multiple Analog Lines

When an application grows beyond 8 lines or so, things begin to change regarding connection to the phone network. First of all, there won't be enough RJ-11 type wall jacks. Also, voice boards with analog loop start interfaces are available up to 16 lines in a single PC slot card. These large voice boards don't have RJ-11 or 14 connectors because they wouldn't fit out the back of the computer.

The solution is to wire directly up to the junction box. In a business environment, that junction box is referred to as a *punch-down block* (also known as a 66-block, Bix block, terminating block, quick connect, cross-connect and patch panel). The punch-down block typically has a larger, 50 pin connector entering the side of the box. This connector is officially called an *RJ-21x* (also referred to as an *amphenol*, for the same reason we say xerox and kleenex).

When you purchase a high density voice board (12 to 16 lines), ask the vendor for an RJ-21x adapter. Then have the telecommunications coordinator obtain an RJ-21x cable that is long enough to reach from you CT system to the punch-down block. The individual tip and ring pairs from each line are connected to the phone company within the punch-down block.

The punch-down block was designed for ease of use. The tip and ring pairs are arranged in an obvious manner. Using a punch-down tool, the wires can be connected without stripping the insulation. If this sounds confusing, just make sure that the telecom coordinator understands what you are trying to accomplish, and they will know how to make the connections.

Connecting to a Telephone Switch

You may wish to consider connecting a CT system to your customer's telephone system. This section explains some of the pros/cons and other considerations. First, its helpful to understand the different types of phone systems.

What Type of Phone System Do You Have?

The purpose of business telephone systems is to allow many users to share a common group of phone lines to the outside world. As illustrated in figure 4-8, these common phone lines from the CO, which connect to the front of the telephone system, are referred to as *trunks*, even though they may not be physically any different than regular phone lines.

Your on-site equipment may be referred to as *customer premises equipment* (CPE), another arcane pre-divestiture three letter acronym. The telephones that connect behind your telephone system are called *stations* and the phone lines that run from the switch to stations are sometimes called *extensions*.

You typically purchase business telephone systems from an *Interconnect* - a telecommunications systems dealer.

44

Business Telephone System Terminology

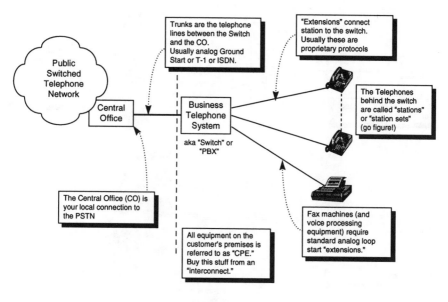

Figure 4-8

Key Systems

In a key system, each of the outside lines are routed to every telephone station. A button, or key, appears on the phone for every outside line. The keys are typically illuminated to indicate whether they are in use or perhaps blinking if on hold. To make an outside call, you press a key that is unused, and the phone is now connected to the outside line. At that point, everything works just as it does on a loop start line. Key systems are typically used in business that have less than 100 stations. See figure 4-9.

PBX vs. Key System

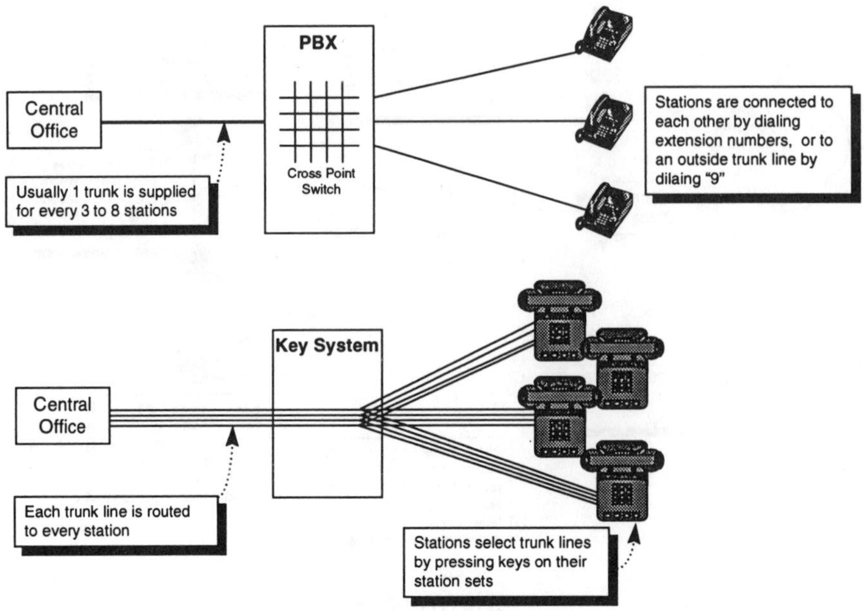

Figure 4-9

PBX

In a *PBX* (Private Branch eXchange), a switch is placed between the outside trunk lines and the in-house station lines. To each station, the PBX looks like a mini-central office. A request for service is initiated by lifting the handset, the PBX provides dial tone. If the caller wishes to make an outside call, dialing a "9" will instruct the PBX to seize one of the outside trunks. The caller is then connected to the outside world. PBXs allow calls to be made from one station to another, without using an outside trunk. See figure 4-9.

Hybrids

Many modern telephone systems today are *hybrid PBX/Key Systems*. Certain businesses like the features of a key system, and other businesses like the features of a PBX. A hybrid, through software, can be setup to provide either or both. If you press a specific key to get an outside line, then you are using it as a key system. If you press "9" to get an outside line, you are using it as a PBX.

A nice generic term for a business telephone system, be it a key system, PBX, or a hybrid, is a *switch*.

Centrex

Centrex is not really an on-site phone system, but instead uses the central office like a big PBX. Features are controlled by pressing sequences of touchtone. If your customer has a Centrex system, they will need to order additional phone lines for the CT system.

Proprietary Protocol Switches

With a proprietary protocol switch, regardless of whether its analog or digital, the line between the switch and the station is not standard. You cannot unplug a *proprietary telephone set* (fancy feature phone) and plug in a voice board, nor a standard analog telephone for that matter. Even though the connector and wall jack may look standard, they are incompatible. To make matters worse, every switch manufacturer uses their own proprietary scheme. In fact, the scheme may be different between two different models from the same switch vendors.

This is a closed architecture strategy that switch manufacturers use to sell proprietary feature phones as unconscionable mark-ups. Don't play that game!

Fortunately, every switch vendor offers an option - plain old analog loop start lines. Thanks to fax machines, which faced the same challenge as CT systems, it is easy to obtain a loop start connection to your customer's switch. Switches typically provide this feature through the addition of a special card which plugs into the switch. You'll need to ask the switch administrator or dealer to provide analog lines for the PBX.

Connecting a CT System Behind the Switch

There are several advantages to connecting your CT system behind a switch. First, callers who need assistance can be easily routed to someone at the site who can personally answer the call. A voice board can be programmed to do this using a hookflash transfer. Second is the economies of scale. The phone lines you need for the CT system can be shared with existing lines or added to the existing lines to justify a new digital service.

You will need to have the switch administrator designate the appropriate number of extensions required by the CT system. This may require purchasing additional line or trunk cards for the switch from the switch vendor and ordering additional trunks from the telephone company.

The switch administrator should organize the extensions for the CT system in a *hunt group* which corresponds to the phone number that will be published for the CT system. A hunt group is a series of extensions organized in such a way that if the first line is busy, the next line is hunted and so on until a free line is found. The switch administrator can also coordinate pulling cable and installing the appropriate connector.

Loop Start Lines in the Sample Application

Dialogic Voice Boards

The Dialogic D/41 four port voice processing board (see figure 4-10) is the workhorse of the CT industry.

Dialogic D/41 4 Port Board

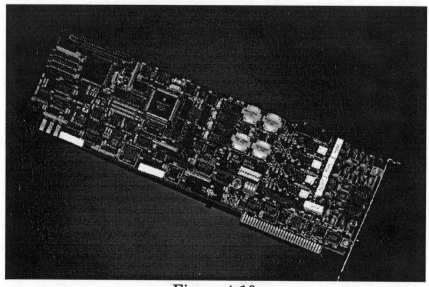

Figure 4-10

The D/41 provides four telephone line interface circuits approved for direct connection to analog loop start lines. It is also available in a half card format.

CT System Configuration with Analog Phone Lines

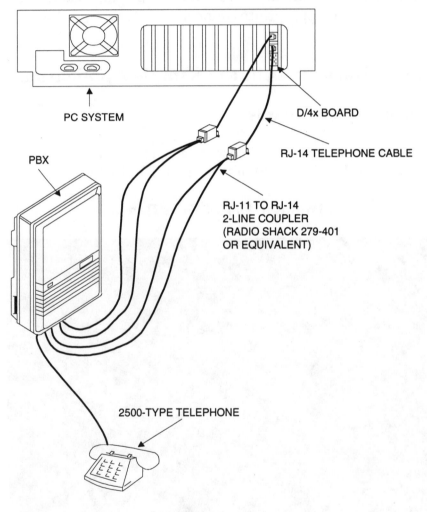

PC SYSTEM

D/4x BOARD

PBX

RJ-14 TELEPHONE CABLE

RJ-11 TO RJ-14
2-LINE COUPLER
(RADIO SHACK 279-401
OR EQUIVALENT)

2500-TYPE TELEPHONE

Figure 4-11

A unique dual-processor architecture comprising a DSP (Digital Signal Processor) and a general purpose microprocessor, handles all telephony signaling and performs touchtone and audio/voice processing tasks. One or multiple D/41 boards can be installed in a single PC chassis (see figure 4-11). The D/41 is only one board in a large computer telephony product line from Dialogic Corporation.

Expert Systems' EASE CT Development Software

Expert Systems is the leading provider of Interactive Voice/Fax Response software.

EASE CT Development Environment

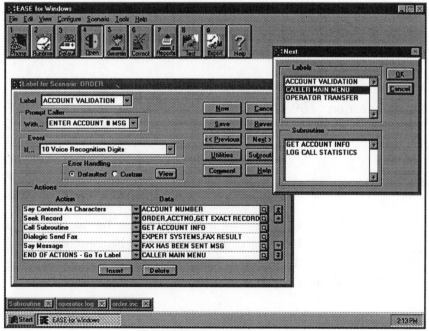

Figure 4-12

51

The company's flagship product, EASE, is a powerful Integrated Development Environment (IDE) for voice/fax processing. EASE walks developers through the logic of creating computer telephony applications with an intuitive user interface. See figure 4-12.

EASE is used by hundreds of OEMs, VARs, system integrators and in-house development teams to create custom applications like Interactive Voice Response, Fax Response, and Application Controlled Switching.

EASE is available for protected mode DOS, Windows 3.1 and Windows 95. EASE for Windows NT is scheduled for the second quarter of 1996.

Expert Systems was founded in 1984, and has been an Open Toolkit Developer for Dialogic Corporation since 1989 and is a founding member of Dialogic's CT Select program. EASE software products are currently deployed in over 40 countries.

EASE has won numerous honors over the years including being named 1995 Product of the Year by Computer Telephony Magazine.

EASE Phone Configuration Screen

Figure 4-13 illustrates how simple it can be to set up a loop start connection behind a PBX. In the bottom right, the telephone network interface is identified as analog with the click of a radio button.

In the bottom left of figure 4-13, a Panasonic PBX is selected from a pop-up list much the way a particular printer is selected under Windows. EASE automatically fills in the default parameters, such as switchhook flash duration, when a particular switch is selected.

EASE Analog Phone Interface

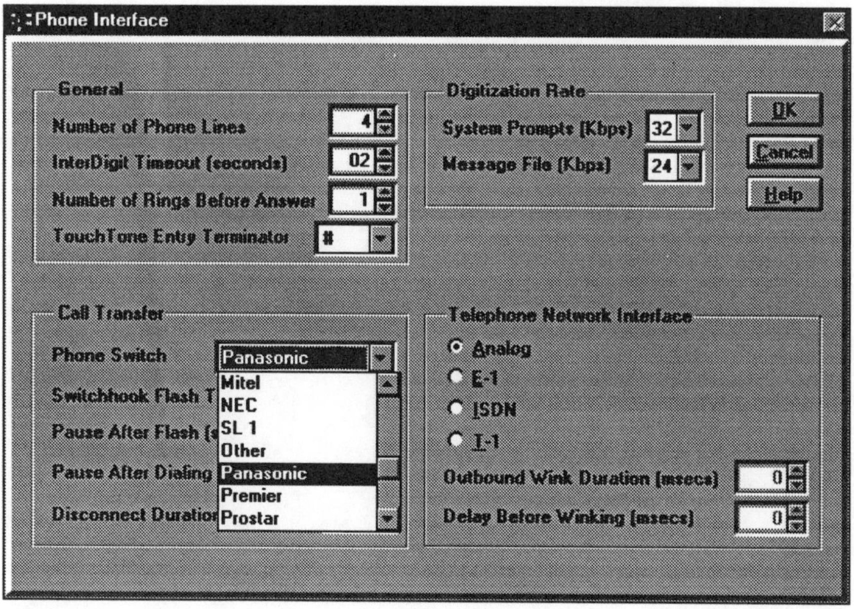

Figure 4-13

In the top left of figure 4-13, the CT system is identified as have four phone lines and callers will be given up to two seconds between touchtone entries. The system will answer calls on the first ring and callers may terminate touchtone entries with a "#" if they do not want to wait two seconds for the system to assume that an entry is complete.

The top right area of figure 4-13 deals with digitized speech which is covered in Chapter 12.

5|
Connecting to T-1 Circuits

Going Digital

Benefits of Digital Telephone Circuits

At some point it is no longer cost effective to connect additional analog lines. Telephone companies will offer a better service deal on a *digital* trunk. Fewer wires will save you installation time and maintenance headaches.

Digital services utilize more advanced technologies to pack up to 30 phone conversations onto two pairs of wires which normally would carry two conversations using analog service. The additional hardware that is required at each end of the wires is offset by the reduced needs for wires and connectors which require manual labor to string on poles and put in trenches. Digital services also offer higher quality voice transmission, just as CD's offer better sound quality than vinyl records.

Digital service also does a better job of providing enhanced features such as the number from which the person is calling (ANI) and which of your incoming numbers was called (DNIS). See Chapter 8 for details on ANI and DNIS. If you're not interested in these features, you'll find that digital services are no more complex to utilize than plain old analog. In fact, many of the signaling schemes on digital trunks emulate analog protocols.

One benefit you can easily take advantage of on digital services is positive disconnect supervision. When the party that calls you hangs up you'll know it immediately.

Where is the cross over point where digital is more cost effective than analog? That varies from country to country, depending on how digital service is *tarriffed* relative to analog. Typically, the cross over is somewhere around 12 lines and is dropping.

Some examples of digital services that we will discuss are T-1, E-1 and ISDN. But first, it is helpful to understand the differences between analog (such as loop start) vs. digital.

Analog Versus Digital Transmission

Voice, video or data signals can be sent using analog or digital transmission techniques. *Analog transmission* uses a continuously varying electro-magnetic wave to represent the information being sent. An analog signal appears on an oscilloscope as a series of sine waves of varying amplitude (voltage) and frequency (cycles per second). In voice transmission, the analog signal is "analogous" to the voice of the person speaking into the telephone.

Digital transmission employs discrete states (0s and 1s) instead of a continuously varying wave. These states are transmitted by varying an electrical current between two pre-established values.

Even though you are using an analog loop start phone line, it is likely that your call is converted to digital for transmission through the public telephone network. The conversion takes place at the phone company central office (CO).

There are a number of methods for converting analog signals to digital format; the most common is called *Pulse Code Modulation* (PCM). PCM conversion is normally a two step process. First, the incoming analog signal is sampled at a periodic rate - 8000 times per second is what is used in the telephone network. Each sample generates a pulse whose voltage is the same as the analog signal's at that point (*Pulse Amplitude Modulation or PAM*). Second, the height of each sample pulse is given a digital value comprised of 8 bits (digital *encoding*). This value represents the voltage of the pulse and thus the analog input at the time of the sample.

These digits are transmitted as a stream of binary bits. When the bits are received at the far end of the circuit, they are used to generate PAM pulses which can be used to reconstruct the original analog signal. Using PCM, a single telephone conversation requires a transmission rate of 64,000 bits per second (8000 samples/sec x 8 bits for each sample). A device called a *codec* (coder-decoder) is used to convert analog signals to digital and back again to analog at the receiving end.

In analog to digital conversion there is a tradeoff between the sound quality that can be obtained and the number of sample bits that must be transmitted. The problem arises when measuring the volume of PAM pulses in order to assign them digital values. Digital encoding requires discrete values, like the markings on a ruler, and the PAM pulses may fall between these values. If the pulse heights have to be approximated, distortion results (quantizing noise) which can be detected by the caller. If finer degrees of measurement are used, more bits must be transmitted to encode them.

To cover the full range of the human voice, 12-bit samples would be needed and a transmission rate of 96,000 bits per second -- far too costly a proposition at this point.

Engineers have solved the problem by using a *non-linear rule* (Figure 5-1) which provides finer measurement at lower and normal volume levels and cruder measurement at higher volume levels. In this way, the quantizing noise is minimized at lower volumes but allowed to rise at higher ones. A non-linear voice encoder serves to compress the voice signal so that it requires less bandwidth for transmission.

The signal can be expanded back to a linear state on the receiving end. This process is called *companding* (compressing-expanding).

The PCM coding and companding scheme used on public networks in North America and Japan is referred to as *Mu-Law*. The format used in Europe is called *A-Law*. A-law is not compatible with Mu-law, but offers comparable performance. Voice boards with digital network interfaces will handle the appropriate coding law.

Non-Linear PCM Coding

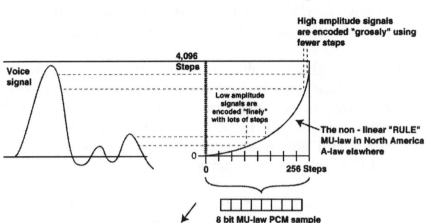

Figure 5-1

The voice quality performance of 64 kilobit per second (kbps) companded PCM has set a worldwide standard, often referred to as toll quality voice. The major factor governing the quality of the audio over a phone line relative to an FM radio or a compact disk, is the frequency bandwidth (maximum cycles per second). The telephone network, with PCM coding, can only pass a maximum frequency of 4 kHz (4 thousand cycles per second), whereas FM radio transmits about 11 kHz, and a CD is 22 kHz.

The 4 kHz bandwidth of the telephone network is due to two factors:

- The 8 kHz sampling rate which was selected as an appropriate tradeoff between data rate and human speech reproduction.

59

- The application of a communications formula called the *Nyquist theorem* which states that two samples per cycle is sufficient to characterize an analog signal.

Given the 8 kHz sampling rate selection, the maximum frequency that can be accurately passed through this communications channel can be no more than 1/2 the sampling rate or 4 kHz of bandwidth. Otherwise, the encoder would not be assured of at least 2 samples per cycle at the highest frequencies.

The frequencies generated in typical human speech are between 300 and 3,000 Hz, sometimes referred to as the *voice band*. Music can be transmitted through the telephone network, but much of the higher frequencies are lost due to insufficient samples per cycle.

There have been recent attempts to improve the quality of the voice through the network, but these involve more marketing hype than technology. For example, if you had a complete end-to-end digital connection, then PCM may have enough dynamic range to here a pin drop. One technique to improve the voice quality is to boost the low frequency and high frequency amplitude, just like a stereo equalizer. This might make Whitney Houston sound a little better over the phone.

T-1 Circuits

T-1 is a popular digital service which has been used in North America since the 1960s. T-1 uses two pairs of wires. Each of the pairs are twisted - a trick which reduces their susceptibility to electromagnetic interference. Each of the pairs carries only one half of the conversation. Splitting the job into transmit and receive paths, and using twisted pairing enables the wires to carry a relatively high speed stream of digital bits - 1.544 million bits in a second (Mbps).

Converting Analog Lines into T-1 Trunks

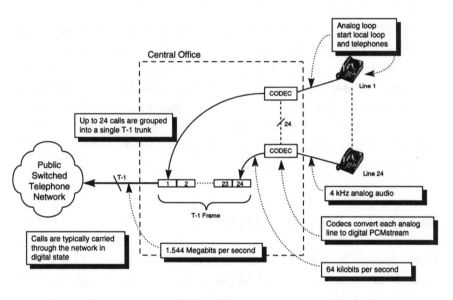

Figure 5-2

As illustrated in figure 5-2, this is enough bandwidth to carry 24 simultaneous voice conversations, each digitized at 64,000 bits per second (bps). An additional 8,000 bps are required to synchronize the multiplexers at each end of the circuit and for signaling. 64,000 bps x 24 channels + 8,000 bps = 1,544,000 bps.

Time Division Multiplexing

A process called *time division multiplexing* (TDM) is used to get these 24 voice channels on the same wire. Each 8 bit sample (see Pulse Code Modulation) from each of the 24 channels is sent down the wire one after another like train cars. Each of the 8 bit train cars is referred to as a *time-slot*.

After 192 bits have been sent (24 channel x 8 bits per channel) a single *framing bit* is added to mark the end of each *frame*. The framing bit helps the receiving equipment determine which 8-bit sample is associated with each of the 24 channels so the samples from each conversation can be rejoined at the receiving end of the circuit. The entire frame process is repeated eight thousand times per second.

Robbed Bit Signaling Over T-1

How are calls placed and received over T-1? The same signaling information that is communicated using AC, DC, and tone signaling on analog circuits can be communicated using digital techniques. There are a number of digital methods but the principles involved are the same for all of them. T-1 uses a method called *A & B bit signaling*. It works in the following way.

Out of each digital bit stream which comprise a voice conversation, two bits called A and B bits are used to transmit signaling information. They are "robbed" from the other bits which convey voice information which is why this technique is often called *robbed bit signaling*.

See figure 5-3. Because these bits are not robbed from every 8-bit sample, but only from every 6th frame, the sound quality of the connection is not affected.

Robbed Bit Format

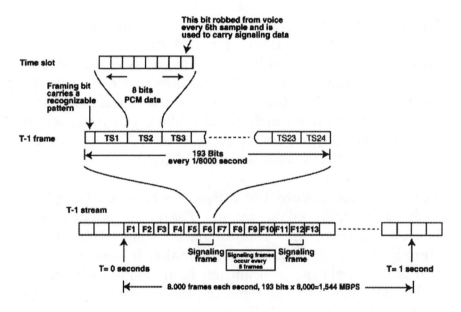

Figure 5-3

Each bit can be set to a value of either 0 or 1 and the four possible combinations (00, 01, 10, 11) can be used to reflect various signaling states. For example, AB=11 can signal that loop current is flowing (device is off-hook) while AB=00 can signal no loop current (device is on-hook).

Pulse dialing can be accomplished by turning the A and B bits off and on just as DC signaling turns loop current off and on to indicate dialed numbers. Tone dialing, which is much faster, is performed using the same DTMF tones as analog phones. It is interesting to note, that although digital T-1 trunks offer greater sophistication regarding voice transmission, the signaling protocols merely emulate those of analog loop start signaling.

On telephone calls transmitted via a T-1 digital channel, a full talk path is allocated as 64 kilobits per second. However, due to robbed bit signaling, a bit is robbed from every 6th frame. This makes the effective data rate for audio 62.667 Kpbs. However, for data transmission, since you cannot be sure which bit is robbed, you must ignore the least significant bit on every frame. This makes the maximum data transmission rate 56 Kpbs.

People cannot discern the difference between 64 kilobit audio and 62.667 kilobit audio. Therefore, the talk path is not significantly compromised when robbed bit signaling is used. This technique makes it easy to route a call as its signaling information is embedded within the talk path. No external signaling mechanisms need to be used, such as with the ISDN protocol.

T-1 Features

In addition to higher channel capacity, T-1 offers some attractive improvements over analog loop start lines:

- With a straight digital connection into your voice board, T-1 results in superior voice quality.

- T-1 provides positive disconnect supervision.

- Services are available over T-1 which can provide you with the number of who's calling (ANI) and which number they dialed (DNIS). See Chapter 8.

T-1 Circuits in the Order Status Application

Dialogic Voice Boards with T-1 Interface

Voice boards are available today, such as Dialogic's D/240SC-T1 (see figure 5-4) which offer an on-board T-1 interface as well as voice processing for all 24 lines.

Dialogic D/240SC-T1

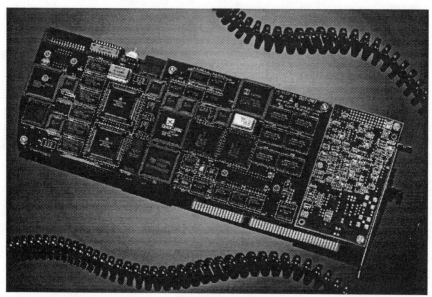

Figure 5-4

Yes, this is a single slot card with a 24 channel network interface that provides command and control of the phone lines, generates voice, and detects touchtones on all 24 lines.

Technically, the T-1 line coming out of the voice board is called a *DSX-1* and has a maximum distance of 655 feet. DSX-1 is suitable for connection to your on premises phone system. If the T-1 line is going outside your building to the phone company central office, its likely you'll be connected through a *Channel Service Unit* or CSU. DSX-1 and T-1 lines are connected using a connector called *RJ-48C*, which looks like a fat RJ-11 modular plug.

EASE T-1 Interface

As figure 5-5 illustrates, changing the order status application to T-1 with EASE is as simple as:

- Clicking the T-1 radio button in the bottom right section and adjusting the wink duration and delay if necessary.

- Increasing the number of phone lines to 24 in the top left section.

- Selecting Other as the phone switch in the bottom left section. The call transfer default settings are automatically updated.

Since robbed bit signaling on T-1 emulates analog signaling, no application changes are required unless you want to take advantage of advanced services such as ANI or DNIS.

EASE T-1 Interface

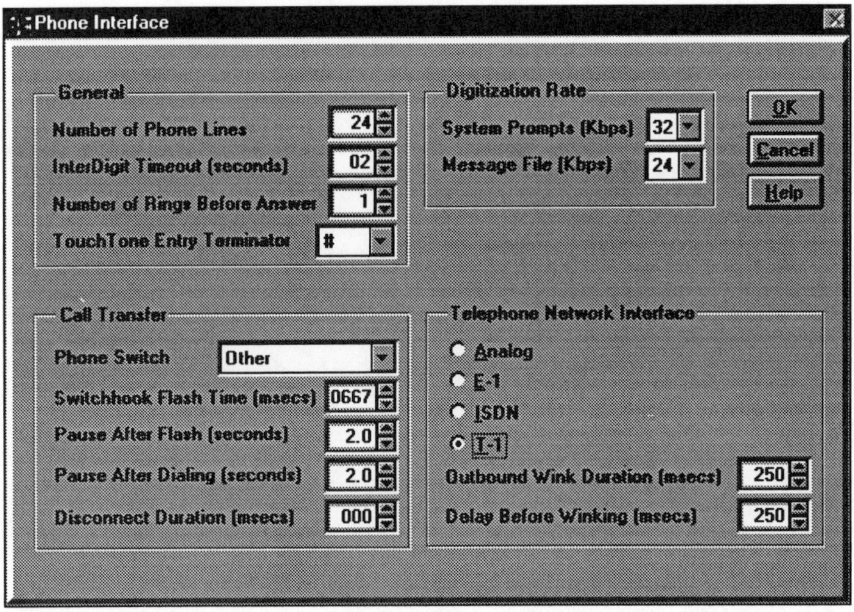

Figure 5-5

6

Connecting to ISDN Primary Rate Interface

ISDN - Integrated Services Digital Network

National *ISDN*, as it is known in the US, utilizes T-1 digital telephone technology and provides a dynamic, highly configurable T-1 connection.

Since T-1 is a common method of carrying 24 telephone circuits, many wonder about the uses for ISDN, especially when they learn that ISDN signaling requires an entire voice channel, reducing the T-1 capacity from 24 voice channels to 23.

But the popular T-1 mechanism of robbed bit signaling has serious limitations. Robbed bit signaling typically uses bits known as the *A and B bits*. These bits are sent by each side of a T-1 termination and are buried in the voice data of each voice channel in the T-1 circuit. Hence the term robbed bit, as the bits are stolen from the voice data.

Since the bits are stolen so infrequently, the voice quality is not significantly compromised. But the available signaling combinations are limited to ringing, hang up, wink, flash, and pulse digit dialing. In fact, the limitations are obvious when one considers Dialed Number Identification Service and Automatic Number

Identification (see Chapter 8) information is sent as in-band DTMF or MF tones. This also introduces a problem: time.

Each DTMF tone requires at least 100 milliseconds to send, which in a DNIS and ANI situation with 20 DTMF digits will take at least 2 full seconds. There is also a margin for error in transmission or detection, which can result in DNIS or ANI failures. With the explosion of telephone related services, telephone companies are turning to the ISDN to provide the more complicated and exact signaling required for new services.

Primary Rate vs. Basic Rate ISDN

In ISDN, a time slot which carries voice communications, and can also carry data communications, is called a *bearer channel* (*B channel*). The channel which carries the signaling information is called the *data channel* (*D channel*).

National ISDN Primary Rate access in North America provides 23 B channels plus one D channel hence the acronym 23B + D (pronounced "23 B plus D"). ISDN Primary Rate access under the *Euro-ISDN* standard provides 30B+D+Framing. See figure 6-1.

Basic Rate Interface is a version of ISDN that is intended for residential or small office use. BRI consists of two B channels and one D channel, and its lower capacity enables it to be provided over the same tip and ring pair of wires that carry analog service.

ISDN Rates

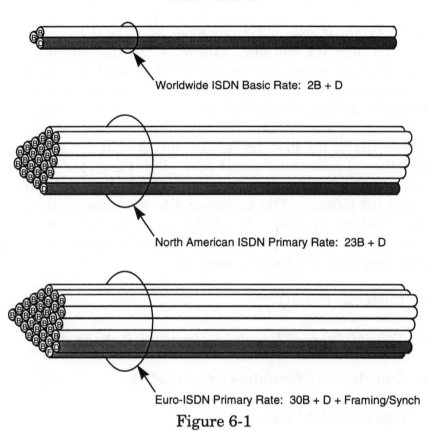

Worldwide ISDN Basic Rate: 2B + D

North American ISDN Primary Rate: 23B + D

Euro-ISDN Primary Rate: 30B + D + Framing/Synch

Figure 6-1

BRI is becoming popular around the world because it offers a high speed data pipe for access to the Internet. The cost of obtaining a BRI line in the U.S. has been dropping, but for basic voice calls and CT systems, BRI is still not a practical replacement for analog loop start lines. This is also true for the rest of the world, with a few exceptions such as Germany, where BRI is replacing loop start for even basic residential service.

ISDN Signaling

ISDN typically uses a T-1 circuit as 23 voice channels and 1 signaling/data channel. The data channel carries the signaling information at a rate of 64 kilobits per second. This speed is many times greater than some of the most powerful modems available.

The actual data that travels on the D channel is much like that of a common serial port. Bytes are loaded from the network and transmitted out to the customer site in a serial bit stream. The customer site then responds with its serial bit stream. The following data items are an example of a data packet sent from the network to a customer site to indicate a new call:

- Customer Site ID

- Type of Channel Required (Usually a B channel)

- Call Handle (Not unlike a file handle)

- ANI and DNIS Information

- Channel Number Requested

- Request for a Response

This packet is responded to by the customer site with a packet that contains:

- Network ID

- Channel Type: OK

- Call Handle

The packets change as the state of the call changes, and finally terminate with one side or the other sending a disconnect notice.

The important concept here is that the information on the D channel could actually be anything - any kind of serial data. So with that in mind, consider the Channel Number Requested field in the packet above. This is the networks' selected channel for the customer site to use.

Because of this high speed data signaling, telephone calls can be placed more quickly, and because of the protocol used, DNIS or ANI transmission failures are impossible. Additionally, since bits are not robbed from the voice channels, the voice quality is better than that of robbed bit signaling.

ISDN Features

ISDN's ability to communicate directly to the phone company's computer enables some interesting features that a CT system may use.

- **Speed:** An outgoing ISDN call is not dialed. A packet is sent and the call is placed "instantaneously," relative to a call placed by tone dialing. Remember that even T-1 robbed bit digital circuits still uses tone dialing. Likewise, a disconnect message, indicating the other party has hung up, is faster. This speed can be important to systems which are handling large volumes of in-bound or out-bound calls. Each line can handle more calls, thus fewer lines are needed. Also, ANI and DNIS information (see Chapter 8) is provided in milliseconds, instead of seconds.

- **Vari-A-Bill:** With AT&T's *Vari-A-Bill* service, a CT system can vary the billing rate of a 900 call at any time during the call. Callers typically select services from a voice-automated menu, with each service individually priced. Contact AT&T for more information.

- **Call by Call Service Selection:** Today's modems are limited to about 28.8 kbps transmission rates because they must treat the network as an analog trunk. Faster digital service is available, but typically only when a permanent circuit or tie line is used. ISDN enables data to take full advantage of the speed of the digital telephone network. When placing a call, an ISDN user can specify that the call is data, not voice. The network will ensure an end-to-end digital connection is established - enabling high speed data transmission. This scheme is especially popular for video conferencing applications.

ISDN in the Order Status Application

Dianatel EA24 T-1/ISDN Primary Rate Interface

The Dianatel EA24 interfaces to a T-1 network using either 24 channel robbed bit signaling or ISDN Primary Rate interface (23B+D). The interface supports a 24 channel PEB connection to Dialogic voice processing boards such as the D/240SC 24 channel voice board or two D/121B 12 channel voice boards.

This is an excellent example of standards based inter-operability between CT component manufacturers. The EA24 can be programmed to alter the signaling information or to pass the data through unchanged to other PC-based interface and resource boards. It can also be used to monitor the signaling data and/or provide signaling protocol such as 'wink' or 'flash'.

The EA24 has been widely deployed throughout the US and Canada and is conformance tested with all major AT&T ISDN services.

Dialogic D/240SC-T1

The Dialogic D/240SC-T1 board (profiled in Chapter 4) provides T-1/DSX-1 service termination and voice/call processing for up to 24 channels...all in a single slot board. New downloaded firmware and a new consolidated API supports ISDN Primary Rate Interface.

EASE ISDN Interface

As figure 6-2 illustrates, EASE provides command and control of ISDN circuits. This screen shows the order status application setting up to take a call on an ISDN circuit. The pop-up list shows the developer all valid options at a particular point in the development process. When an Action is selected, dialog boxes guide you through the details. Of course, context sensitive on-line help is available.

EASE ISDN Command and Control

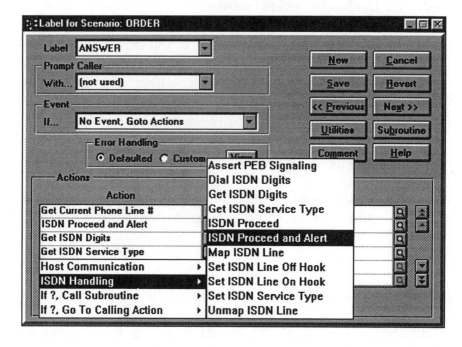

Figure 6-2

7

Connecting to E-1/Euro ISDN Circuits

E-1 Circuits

The European cable standard that is roughly equivalent to T-1 is referred to as *E-1*. E-1 transmits at a rate of 2.048 Mbps (vs. 1.544 Mbps for T-1) and allows for 32 *time slots* (vs. 24 for T-1). 30 of these time slots are used to carry voice conversations. The remaining two are used for framing and signaling information. Unlike robbed bit T-1, which robs a few bits from each voice channel, E-1 robs an entire pair of time slots, and uses them for framing and signaling instead of transmitting voice. This scheme is referred to as Channel Associated Signaling (CAS). Numerous E-1 protocols and ISDN variants are available over E-1 circuits.

E-1/Euro ISDN Protocols

From a protocol viewpoint, E-1/Euro SDN is more a state of mind than a standard. Generally speaking, each country uses one or more specific E-1/Euro ISDN protocol. The differences depend on the type of central office switch being used and the version of software. You must be aware of what type of switch, and subsequently what type of protocol is available. Country specific protocol information must be specified when ordering E-1 interface components.

The following is a list of country specific protocols supported by the Aculab Digital Network Interface card:

Country	Protocol
Australia	TS014
Australia	P2/R2D(CAS
Australia	P2/TS003/TPH1271
Austria	FZA Dbb III 0215
Belgium	National R2/DTMF
Brazil	MFC R2
CCITT (R2)	Q421
CCITT (R2)	Q421/Q441 A5/A9 CLI
Chile	MFC R2
China (#1)	MFC R2
Columbia	MFC R2
Croatia	MFC R2
Euro ISDN	ETS300-102
France	VN3/VN4
France	MF Socotel (R1)
Germany	FTZ 1TR6
Greece	OTE 2bit CAS
Holland	ALS70D (T11-53E)
Hong Kong	CR13 IDA-P
Indonesia	SMFCR2
Israel (R2)	MFC R2
Malaysia	MFC R2
Mexico	MFC R2
New Zealand	TNA 133
Norway	MFC R2
Pan European Euro ISDN	ETS1/ETS300/102
Peru	MFC R2
Portugal	Decadic R2

Portugal	MFC R2
Singapore	FETEX 150
South Africa	MFC R2
Spain	MF Socotel (R1)
Sweden	P7
Sweden	P8
Switzerland	SwissNet2
Thailand	MFC R2
UK	DASS-2
UK	DPNSS
UK	DPNSS, enhanced
UK	SS DC5A
UK (BT CAS)	BT Callstream SIN205
UK (MCL CAS)	OTR001 PDI DDI
UK (MCL CAS)	OTR001 PDI non-DDI
USA	AT&T TR41449/41459

E-1/Euro ISDN in the Sample Application

Connecting to E-1/Euro ISDN Circuits

E-1 service is offered using RJ-48C connectors, which looks like a fat RJ-11 modular plug, as well as through a pair of 75 ohm coax cables. When you order your voice board, be sure you know what type of connector you'll need.

Aculab E1/ISDN Digital Network Cards

Aculab E1/ISDN Digital Network Cards provide single (30 channel) or dual (60 channel) E-1 service termination and time slot switching. Each E-1 port has its own processor to handle call control functions.

Aculab has considerable experience with country specific protocols (see above table) which are provided as downloaded firmware. The Aculab card connects via PEB to a Dialogic D/320SC card which provides 30 channels of voice processing. This is another excellent example of standards based inter-operability between CT component manufacturers.

Dialogic D/300SC-E1

Provides E-1 service termination and voice/call processing for up to 30 channels, all in a single slot board. New downloaded protocol specific firmware, new on-board call control processing and a new consolidated API will support E-1/Euro-ISDN interfaces.

EASE E-1/Euro-ISDN Interface

As figure 7-1 illustrates, EASE also provides command and control of E-1/Euro-ISDN circuits. This screen shows the order status application setting up to answer a call on an E-1 circuit. The pop-up list shows the developer all valid options at a particular point in the development process. When an Action is selected, dialog boxes guide you through the details. Of course, context sensitive on-line help is available.

EASE E-1 Command and Control

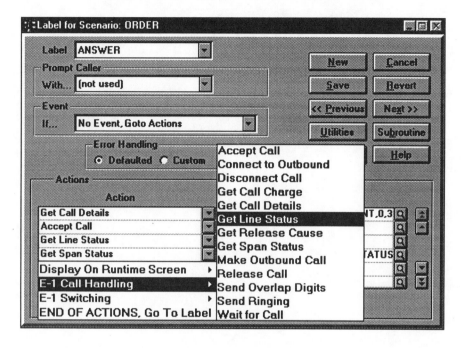

Figure 7-1

III|
Caller Input

8|
Caller and Called Party Identification

9|
Touchtone Detection

10|
Dial Pulse Detection

11|
Automatic Speech Recognition

8|
Caller and Called Party Identification

Knowing Who's Calling

Automatic Number Identification

Even before you answer an incoming telephone call, there is some information available about that caller. First, it is possible to obtain the telephone number from which the person is calling. The telecommunications term for this is *Automatic Number Identification* (ANI). *Caller ID* is a type of ANI service available over loop start lines.

For CT systems, the ANI information can be used to provide faster service. The telephone number associated with an incoming call can be sent back to the host computer, where a database lookup for a pre-existing account can commence even before the call is answered. Of course, verification is still necessary by asking the caller to enter their account number, using touchtone, dial pulse, or speech recognition. But in the case where the caller is calling from their own telephone, verification and access to data can be much faster.

ANI can also be used to override a callers request. For example, a caller could call in for technical support; but, an ANI database lookup indicates that this caller is not covered by a current support plan. In this case the caller

85

could be transferred to sales or accounting regardless of what menu selection they might make.

In the case where the customer is calling from a different phone or an office phone not yet in the database, the ANI may not be recognized and normal caller input will be required. Of course, this new ANI phone number could be added to their customer record after the customer has been identified through touchtone, dial pulse, or speech recognition.

Caller ID

Caller ID is an ANI service available on loop start phone lines. The number (and name of the caller, in some states) is sent to the phone or voice board using modem signals between the first and the second ring. See figure 8-1.

You have to subscribe to this service from your local phone company, and you'll need a voice board, such as the Dialogic D/41E-SC which supports the Caller ID feature.

Caller ID has been available for quite some time, but was marginally useful in that the calling number was only presented for local calls. While this is great for a pizza shop, its not very useful for an application servicing a wide geographical area.

As of December 1, 1995, the FCC has mandated that Caller ID must be passed between regional phone companies and the long distance providers. Caller ID in the US is now approaching nationwide number delivery.

How Caller ID Works

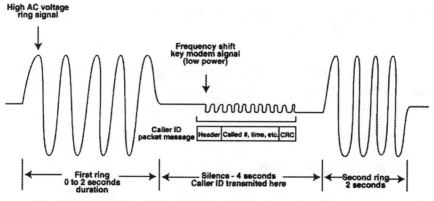

Figure 8-1

Schematic of Telephone with Caller ID

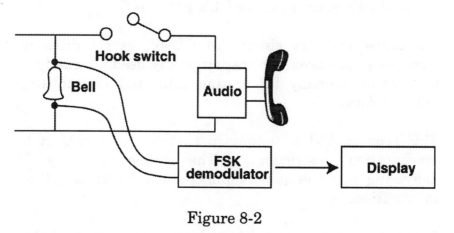

Figure 8-2

Canada and the UK also have Caller ID service available. In the UK, its called CLIP.

Caller ID is also subject to *presentation restriction*. If a caller does not wish the called party to receive their phone number, they can block it by pressing a touchtone sequence such as *67.

Of course the called party can request service that blocks any calls that do not provide Caller ID. See figure 8-2 for a diagram of a telephone equipped with Caller ID.

ANI Over T-1 Service

If you subscribe to an 800 and T-1 service from a long distance providers, you may be able to obtain ANI information.

In this case, the ANI information is sent to you using either standard touchtones (DTMF) or an older scheme call *Multi-Frequency* (MF or R1/MF) tones.

MF tones are very similar to DTMF, except different frequencies are used. MF was originally used internally in the phone company for routing calls and transmitting billing information.

This type of ANI is unique in that it is not subject to presentation restriction. The phone company is delivering you the same number used for their billing information.

There is no way for a caller to block your receipt of this information. Thus, the FCC has recently placed restrictions on the use of the numbers and information you collect, especially in the area of selling this information to others.

The ANI digits are communicated from the telephone company central office (CO) to a CT system using a variety of protocols, the most popular is called *wink start.* When the CO has an incoming call, it signals using the A and B signaling bits in the "incoming call state." The CT system responds with a brief change of the A and B signaling bits.

This handshaking signal is called a *wink.* The CO responds to the wink by transmitting the ANI information using DTMF or MF, or perhaps even signaling pulses. When complete, the CT system goes off-hook to accept the call.

Another protocol for receiving ANI information is called *immediate start.* In this situation, the CT system goes off hook, then the CO immediately transmits the ANI digits and connects the incoming call.

ANI Over T-1 in the Order Status Application

Figure 8-3 shows the order status application receiving 10 ANI digits as it answers a call coming over a T-1 circuit.

As the actions indicate, the CT application compares the Caller ANI to a database called BAD ANI. The results of this comparison can be used to deny access to technical support if the ANI lookup determines that the caller's support plan has expired.

ANI with EASE

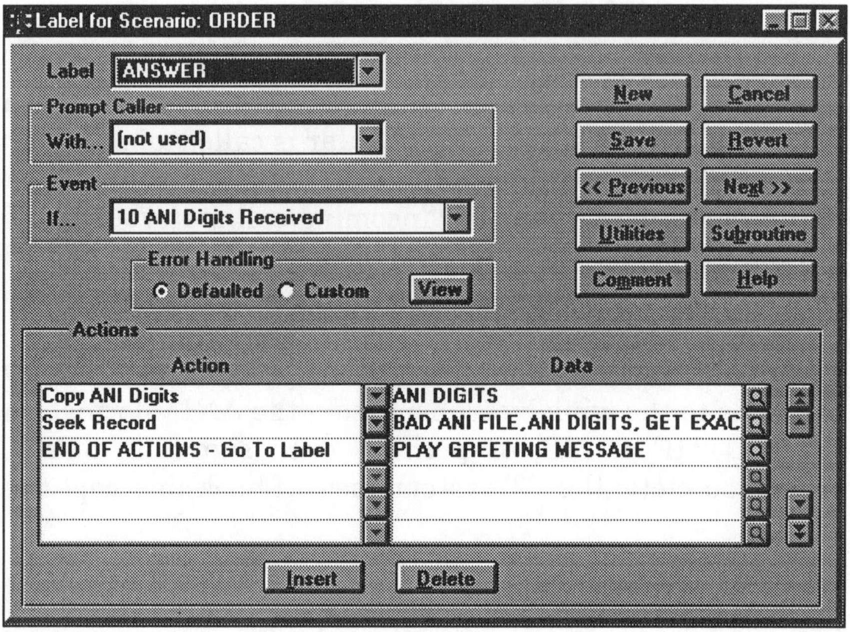

Figure 8-3

ANI Over E-1

ANI is available over E-1 trunks through a complicated tone signaling protocol called *R2 / MF compelled signaling.*

Compelled signaling requires your equipment to respond with an appropriate handshaking tone for every tone that is received from the CO. What makes it complicated is the variety of different protocols that exist, depending on the country and the type of CO switch being used.

ANI Over ISDN

ANI over ISDN is faster than other protocols. Caller ID forces you to wait for several rings. The tone signaling used on T-1/E-1 can take seconds. ANI is provided over ISDN in an instant. In fact, your phone company may charge you for each ANI delivery, and give you the option to decide on a call by call basis whether or not you wish to purchase ANI for that call (usually about 1 penny per call).

In some markets ISDN is the only way that ANI is offered. This is true for Japan, Canada, and in the US for customers of AT&T and the RBOCS.

Knowing Who Was Called

Dialed Number Identification Service

If your CT system is providing more than one type of service, you may wish to consider using a different phone number for each service. Doing so can provide for faster and more dedicated service. But dedicating a separate line for each service can be expensive and inflexible. This is where *Dialed Number Identification Service* (DNIS) comes in. DNIS is a service that is offered over T-1, E-1, and ISDN. Technically, it is handled much the same way as ANI.

The CO will tell you, prior to your acceptance of an incoming call, which of your numbers the caller dialed. This makes it possible for a single CT system to take calls for two different applications, on the same physical equipment. Before a call is answered, the DNIS information is used to start the appropriate application. From that point on, the CT system appears as a dedicated system.

ANI and DNIS over ISDN in the Sample Application

Figure 8-4 shows the order status application receiving 10 ANI and 7 DNIS digits as it answers a call coming over an ISDN circuit.

EASE ANI and DNIS over ISDN

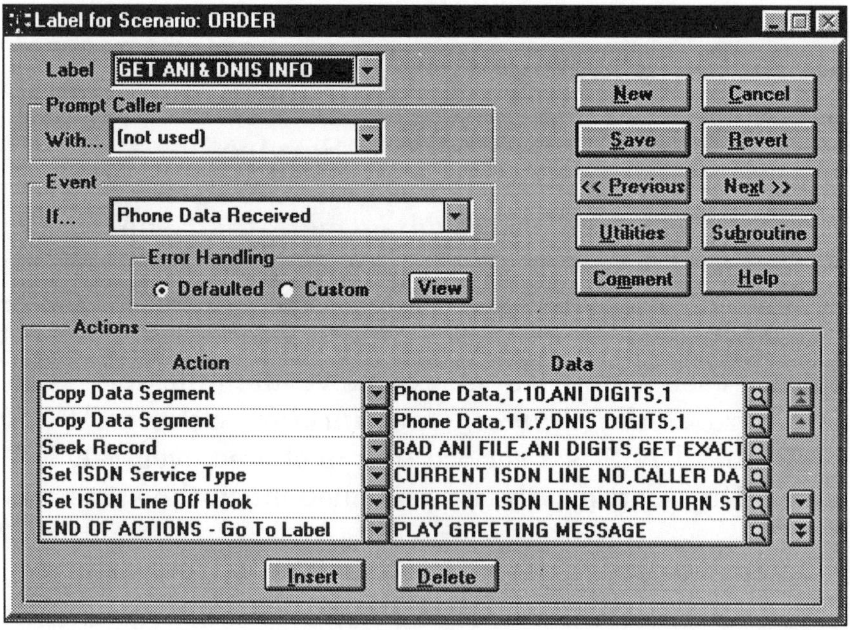

Figure 8-4

In addition to denying access to technical support if the ANI lookup determines that the caller's support plan has expired, the CT application knows that the caller is calling for the order status application and thus plays the appropriate greeting message.

Direct Inward Dialing

Obtaining called party information over analog lines is possible, but a bit more cumbersome. On analog lines, there is a service similar to DNIS called *Direct Inward Dialing* (DID). DID was designed for a different purpose - to allow calls to a business PBX to be routed directly to an extension. On a DID line, the phone company provides you with only the last four digits of the number the caller dialed.

A CT system requires specialized equipment to access DID. Most voice boards with loop start interfaces do not support DID. There are specialized boards available, such as the Dialogic DID/120. There is also equipment available, such as the Exacom DID 200, which sits between the voice board and the CO, receiving the DID information and relaying it to the voice board. The process involves winking and tone/pulse transmission, similar to that described in the T-1 section. In either event, special DID lines are required from the phone company. Typically, you cannot place outbound calls on a DID line.

9

Touchtone Detection

Dual Tone Multi-Frequency (DTMF)

Dual Tone Multi-Frequency (DTMF) is the technical term for the touchtones used on the telephone. DTMF signaling was originated for use in dialing telephone numbers over the Public Switched Telephone Network (PSTN), and has been adopted by the computer telephony industry as a data input method from callers. The international specification for DTMF signals are governed by the *International Telecommunications Union* (ITU), formerly known as the CCITT.

DTMF is the most inexpensive, positive and reliable method for a CT system to request input from callers. The only catch is that in many parts of the world, the typical residence still has a rotary or pulse dial phone.

How DTMF Works

DTMF signals are actually a combination of 2 tones: one low frequency and one high frequency. There are four low frequency tones (697, 770, 852, 941 Hz) and four high frequency tones (1209, 1336, 1477, 1633 Hz). Each of the rows on your touchtone telephone keypad corresponds to a different low tone, and each of the columns corresponds to a high tone. See figure 9-1.

Touchtone Keypad

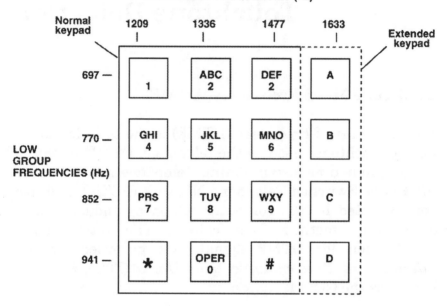

Figure 9-1

Pressing a particular key generates a unique combination of the two tones. 16 combinations are possible. A DTMF entry is often referred to as a *digit*. A fourth column of buttons is defined but not present on your telephone. These buttons are designated as A, B, C and D, and originally had a special use for military communications over the phone network. Today, most voice boards can generate and detect A,B,C and D tones, but they are used mainly for system testing.

As discussed in the digital interface section, there is another type of tone signaling used in the network called MF. DTMF uses a similar coding scheme as MF, however, the fundamental frequencies are different. Since MF was used to transmit billing information, it was necessary to create a different set of frequencies for deployment into the hands of the subscribers.

The Challenges of DTMF Detection

The advent of voice processing and computer telephony presented new challenges to the technology of DTMF detection. These four challenges are:

- **Attenuation:** DTMF signaling has been used since the 1960's as a method for dialing. In the dialing case, the tones are only traveling from a telephone over a single local loop to the CO. In a computer telephony deployment, the tones must travel over a second local loop - between the CO and the CT system - and can therefore be degraded in amplitude which is perceived as a loss of volume or loudness known as attenuation. See figure 9-2.

DTMF Signal Loss Over Network

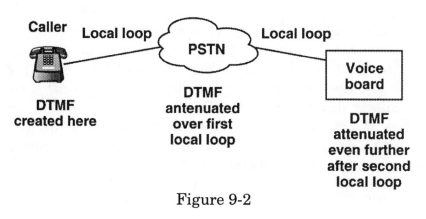

Figure 9-2

- **Power User Input:** A second complication with DTMF detection on a CT system is the "power user" - a frequent system user who expects to enter digits at a speed which is unacceptable to the local central office for outbound dialing. Power user input can be a problem because of either a digit duration that is too short or a duration between digits that is too short. Although some power users adopt the unsecure practice of programming their account number and even password into a speed dialer, this is not a problem for CT systems because the speed dialer maintains the proper durations for digits and between digits.

- **Cut Through:** A CT system user today expects to be able to enter touchtones while a voice prompt is being played by the system. They expect touchtone entries to interrupt the voice prompt and be processed immediately - a situation known as cut through. Cut through is exceptionally challenging for the CT system because not only are the DTMF signals entered by a finger flying power user, they are attenuated by two local loops, and also mixed in with the voice prompt being transmitted by the CT system.

- **Talk Off or Falsing:** It is entirely possible for the prompt voice, during the normal course of speech, to create a tonal pattern that appears like a DTMF. Since the detector is active during this speech, it can falsely detect a DTMF. This is an undesirable event, as it can cause the system to branch to a different state, leaving the caller confused. This situation is called talk off or falsing.

Advent of Digital Signal Processors

Early voice processing board designers eventually found that the inexpensive tone detector chips, designed for use in the CO, were unsuitable for voice processing applications. The standard for voice boards today is to utilize *Digital Signal Processor* (DSP) technology to detect DTMF. DSPs enable not only a more intelligent detector, but they are programmable, so the designer can alter performance if needed to suit a particular environment.

DTMF detection is the single most critical function of the voice board. The following four sections discuss the four challenges of DTMF detection in more detail.

Handling Attenuation

DTMF amplitude is typically specified by a unit of measurement called a dBm for decibels below one milliwatt (mW). It is the output power of a signal referenced against an input signal of 1mW. 1 dBm is an absolute measurement representing 1mW of power over a 600 ohm load.

Never mind that, a good quality voice board must detect DTMF over the range of at least +3 dBm (loudest) to -36 dBm (quietest). The upper or loud specification is important if the voice board happens to be located very close to a CO or PBX. The lower or quiet specification is important for detecting users who have very poor analog local loops at their end. DTMF below -55 dBm should be rejected, otherwise it may detect DTMF interference from an adjacent telephone line (*crosstalk*).

These specifications can vary slightly in different countries. Make sure the voice board you select has a configurable range to suit your market. Check for the above specifications.

One other note: Some voice boards with on-board T-1 and E-1 interfaces specify DTMF amplitude in dBm0, which is the digital equivalent of the dBm. 0 dBm is about 3 dB above dBm0, so the specifications may vary somewhat from what is listed above.

Handling the Power User

The customer is asked to enter their 10 digit account number. They punch it in as fast as they can. Correctly. The system responds with "I'm sorry, that is not a valid account number. Please try again." Poor DTMF detection time characteristics can cause this problem.

DTMF for dialing is specified to be a minimum of 50 milliseconds per digit, with a 50 millisecond pause between digits. This amounts to 10 digits per second. The typical power user can violate this once or twice during a string of entered digits. A good voice board should specify digit detection down to 40 milliseconds or lower (see figure 9-3).

The time between digits is important, too. The voice board should be able to detect two successive *different* digits with no pause. Two successive *like* digits will require at least a 40 millisecond pause, otherwise they may be mistaken for one digit.

Minimum Time for DTMF Detection

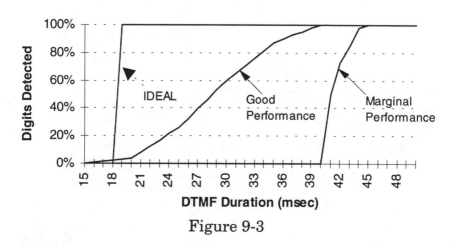

Figure 9-3

Handling Cut Through

Good cut through performance is necessary for repeat caller satisfaction. Cut through allows the caller to enter DTMF digits without waiting for the CT system's voice prompt to finish. To the caller, the DTMF "cuts through" the voice to instruct the system.

As sophisticated as DTMF detection algorithms get these days, they cannot detect DTMF that is mixed in with human speech, unless the speech is significantly quieter than DTMF. Analog loop start lines carry both sides of a conversation on the same pair of wires. When two people talk over a telephone line, they really don't care if their voices mix together. However, a voice board is attempting to play out voice, while at the same time listening for DTMF. Therefore, the playback voice is heard by the DTMF detector.

101

Voice Cut Through Problem & Echo Cancellation

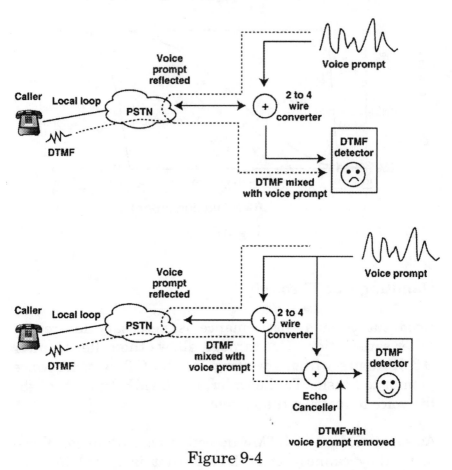

Figure 9-4

Voice boards accomplish cut through by using an echo cancellation algorithm in the DSP (see figure 9-4). Here is how echo cancellation algorithm works. When a voice board begins playing voice, the algorithm looks at the echo that results at its receiver. The algorithm uses the difference between the outgoing signal and the received signal to "model" the telephone line. The playback speech can then be passed through this model, creating an estimate of the echoed signal. The estimated signal is subtracted from the real echo. The resulting received signal contains only information that originated outside the voice board -- DTMF.

DTMF cut through is difficult to measure and there is no standard unit of measure to quantify performance. Dialogic tests all of their voice boards using what they call "the DTMF torture test."

A SAGE Instruments 930A Communications Test Set is used for the test. This device can generate a DTMF tone of any amplitude, duration, frequency variation, etc., all under computer control. Computer control allows automation of the test, ensuring statistically significant results.

The test is performed by playing digits to the voice board while it is playing back a specific voice file. The voice file has all of the silence between the words trimmed out, so the test is tougher. Strings of digits are played to the board, and statistics on the number of digits correctly detected are recorded.

To make the test challenging, a loss prone, reflective, 12,000 foot local loop is used between the voice board and the test equipment. The test can run as many as 10,000 DTMF digits, at varying amplitudes.

The results can be plotted to show relative cut through performance. Dialogic typically runs this test on of its voice boards, as well as other voice boards on the market, to ensure that it is providing the industry's best cut through. See figure 9-5.

DTMF Cut Through Performance

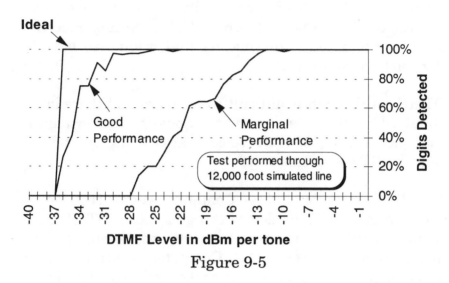

Figure 9-5

Handling Talk Off or Falsing

The voice bandwidth is typically 300-3000 Hz and represents the range of audio frequencies that is essential for transmitting human speech. DTMF signals are in-band signals, which means the DTMF tones are in the voice frequency band and are carried on the same circuit as the voice path.

Talk off is a hazard of in-band signaling. Talk off occurs when frequencies in a speaker's voice (typically a female voice) are mistaken for DTMF. This is most likely to occur with voice messaging applications where the caller's voice is being recorded at the same time that DTMF detection is enabled.

Talk off is one problem that is not new to the telecommunications industry. *Bellcore*, the research arm of the RBOCS, has an established talk off performance test. The test is performed by playing a test tape to the detector (in our case a CT system) and counting the number of times the detector falsely triggers, known as "hits." The fewer hits, the higher the detector's immunity to talk off.

The test tape is challenging - it consists of over 100,000 snippets of human speech, each one chosen for it's particular ability to mimic DTMF. The test takes approximately 3 hours to perform. Acceptable performance by Bellcore's standard is 470 hits (out of 100,000 possible). In the case of a CT system, typical performance should be a maximum of 20 hits and preferably less.

Another test is available from Mitel, a manufacturer of DTMF detector integrated circuits. The Mitel test is much less demanding, consisting of about 20 minutes of human speech. Typical integrated circuit detectors perform with less than one hit on the Mitel tape. This means sometimes they receive no hits, occasionally they receive 1 hit. The Mitel test is no longer an acceptable measure of talk off performance. Look for Bellcore test results.

One variation of talk off that is particularly embarrassing is play off. This occurs when the echo of a voice file being played back causes the board's DTMF detection to falsely detect a touchtone. The echo cancellation on today's DSP based voice boards makes play off almost non-existent,

though. Also, many prompt recording tools offer the
ability to filter DTMF out of your prompts.

DTMF Performance Tradeoffs

The various characteristics of DTMF detection
performance, such as talk off immunity, cut through and
minimum time for detection are interrelated.

DTMF Detection Performance Trade-Off

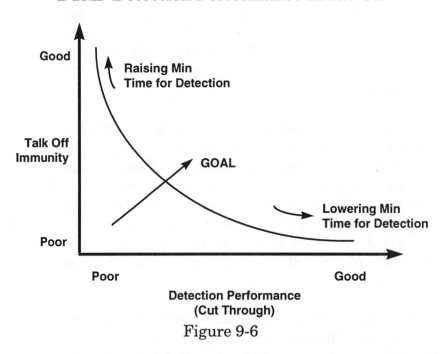

Figure 9-6

While most voice board vendors offer parameters that can
be adjusted to improve performance in a given area, often
this improvement comes at the expense of poorer

performance in another area. For a given detection algorithm, there is a "budget curve."

As figure 9-6 illustrates, raising the minimum time for detection can improve talk off immunity. But this may worsen cut through performance. Be sure that the voice board you purchase not only has acceptable specifications, but that they are tested in harmony.

Caller Input Error

Once you have selected the best available voice processing boards to assure that your CT system can detect touchtones with high reliability, you have another problem related to caller input - and that is caller input error. Up to 60% of your application code usually deals with what the CT system should do when the caller makes an input error. These errors typically fall into one of the following three categories:

- **No Entry/Time-out:** The CT system prompts the caller to enter digits and waits for a specified number of seconds for touchtone entry to start...but nothing happens. Maybe the caller didn't understand the instructions, maybe they didn't have the information available. Who Knows? The bottom line is that the CT system asked for input and got none.

- **Invalid Entry:** The CT system prompts the caller to enter a six digit account number and the caller enters five digits, or seven, or any number other that the requested number. Similarly, the CT system prompts with, "For account inquiry, press 1; or, for order status, press 2." ...and you guessed it, the caller enters 9. These are invalid entry errors.

- **Disconnect:** The CT system is prompting for input, or processing a request, or any of a million other CT functions and the caller simply hangs up the phone in the middle of everything. This is called a disconnect error.

Error Handling

Consistent, automated error handling is a critical success factor in CT system implementation - both from the caller and the developer perspective.

Callers need a consistent safety net that makes them secure while navigating within a CT system. Global error handling can be simply defined and wrapped around every node of logic requesting caller input. A sample of typical global error handling is:

- Caller will be given eight seconds after a prompt has finished playing in which to start making touchtone entries. If no DTMF is received during this time, this will be considered a no entry/time-out.

- Callers who make a no entry/time-out or an invalid entry error, will hear the following: "We cannot identify that entry. Please try again." Then the most recent prompt will be re-played.

- If the caller makes another no entry/time-out or an invalid entry error, they will be transferred to an operator for assistance.

- If the caller disconnects during normal processing, all speech and processing will be terminated and the line will be reset to take another call.

Developers need automation to implement this kind of error handling around every node of logic requesting caller input. Error handling can be 60% of the application code, but it is tedious, mistake prone work if not automated. Plus the developer must have the flexibility to override the global error handling to address specific situations.

Touchtone Input in the Order Status Application

Figure 9-7 shows a node of logic (label) called Enter Order Number. By reading the screen you can see that the caller will hear a prompt called Enter Order Number which we know from the specification says, "Please enter your 5 digit order number now."

By examining figure 9-7, you see that the CT application is looking for an event of 5 touchtone digits received. Under the actions section, you can see that the five digits will be copied from the phone data buffer to a variable called Order Number.

Touchtone entry is the typical way that CT system receive caller input.

EASE Touchtone Input

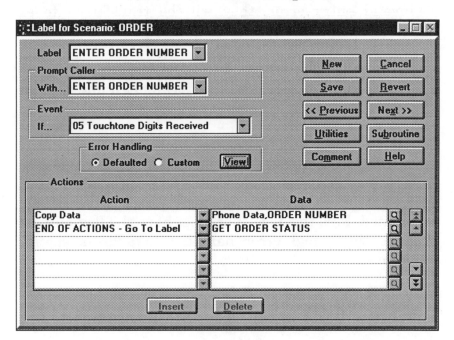

Figure 9-7

Error Handling in the Order Status Application

But what happens if the caller makes a mistake. Not to worry! There is an automatic safety net. Figure 9-8 shows the default error handling for this node of logic which can be seen by clicking the View button.

Reading the error handling dialog box from top to bottom you can see what will happen in various error conditions. If the caller makes a mistake, they will hear a retry message like "We cannot identify this entry. Please try again." Then the Enter Order Number prompt will be repeated. The caller will be given two more attempts and if they are still not successful, processing will branch to a node of logic called Operator Transfer.

EASE Error Handling

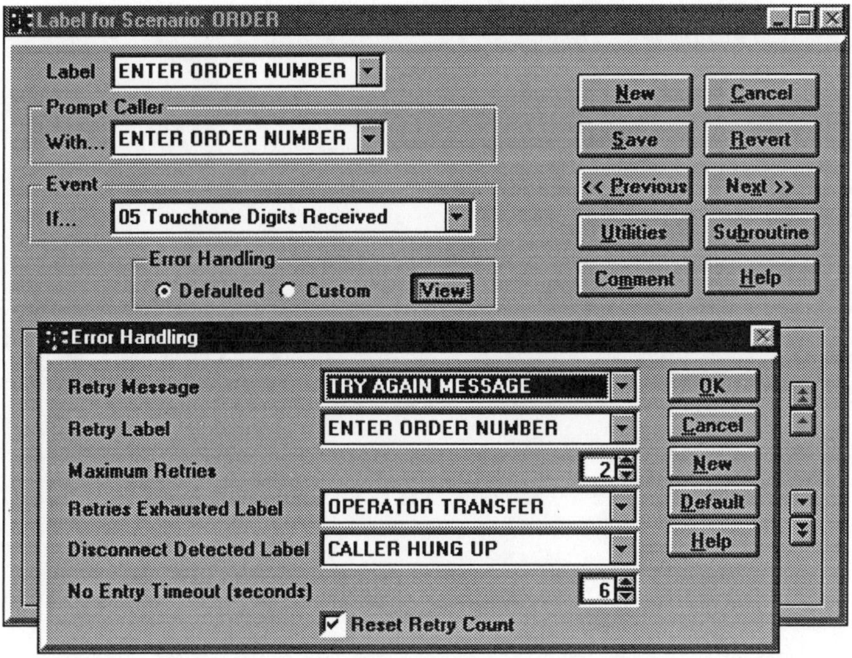

Figure 9-8

If the caller hangs up in the middle of the call, processing will branch to a node of logic called Caller Hung Up. And finally you can see that the caller will be given a maximum of 6 seconds in which to begin entering their order number.

This type of default error handling is automatically wrapped around every node of logic you define. Of course you can override it by simply clicking the Custom button and changing any item you choose.

10

Dial Pulse Detection

Pulse or Rotary Dialing

Pulse dialing or *rotary dialing* is associated with telephones having a circular dial and finger holes corresponding to the digits 1 through 0. The "*" and "#" buttons of a touchtone phone are not represented on a pulse phone.

As the dial returns to its normal position after being turned by a finger in a hole, it opens and closes the electrical loop sent by the telephone company central office momentarily breaking then re-establishing the direct current (DC) circuit. The number of breaks represents the digit dialed. For example, the circuit is broken three times for the digit "3". The phone company central office counts these evenly spaced breaks and determines which digit has been dialed. A person listening to these circuit breaks hears them as clicks.

Dial Pulse Detection

The telephone company central office serving a dial pulse subscriber has direct access to the DC circuit information so that dialing information is received with high reliability. A CT system, on the other hand, is a third party to this process and does not typically have access to this DC circuit information. Dial pulse detection must rely on counting the clicks made each time the DC circuit is broken and then re-established. This may sound simple, but the wide diversity of telephone networks and central office switches makes this a complicated process.

Low Touchtone Availability.

Dial Pulse Detection can provide an immediate, field-proven solution to one of the voice industry's biggest problems in Europe, Asia, South America, and Africa - the lack of dual-tone multi-frequency (DTMF) or touchtone service.

Since touchtone service has been widely available in North America for the last 10-15 years, many voice system suppliers and developers are surprised (and disappointed) to discover that the great majority of the countries in the world are still dominated by pulse dial telephones. This includes Western European countries like Germany (with 90% pulse subscribers), South American countries, let alone other developing countries in Asia.

Despite substantial investments in telecommunication networks worldwide, the expected overall rate of change to touchtone in subscriber's telephones is still relatively low as the existing installed base will remain active and account for the majority of telephones for some years to come. For example, if we look at the USA, even after 20 years of touchtone availability, approximately one quarter of telephone subscribers are still connected to pulse dial lines.

Touchtone/DTMF Penetration Rates

Austria	25%
Australia	30%
Benelux	40%
Denmark	75%
France	50%
Germany	10%
Japan	30%
Italy	10%
United States	70%
New Zealand	30%
Norway	85%
Portugal	10%
South America	10%
Spain	15%
Sweden	85%
Switzerland	30%
United Kingdom	30%

Dial Pulse Detection Versus Voice Recognition

Voice processing suppliers in international markets have started searching for solutions to this problem. After realizing that the network upgrades will not change the situation in the short term, two technologies have arisen as possible solutions - voice recognition and dial pulse detection.

When analyzing the advantages and disadvantages of each technology, it is important to disregard the associated generic claims and perceptions. It is recommended that these technologies are reviewed from the narrow angle of specific application and system needs.

The pros or cons of a certain technology relate directly to the following criteria:

- **Type of Application** - Does it require interactive input? Does it involve string digit input?

- **Nature of Application** - Is it of a personal or discrete nature such as bank inquiry where overheard account numbers and passwords may be an issue?

- **Type of User** - Is it a closed group of users or public and random callers?

- **System Configuration** - Is it compatible with the desired solution (analog or digital environments, and industry standard open protocols?

- **Cost per Line and per System** - What is the cost/performance? Is a shared resource configuration possible and how does this effect the cost/performance.

- **Proven Performance in the Target Market** - Has the solution/technology been proven to perform properly in the market where the system will be operated?

- **Cultural Factors** - Does the target audience have any cultural bias against speaking to a machine?

Quick Guide to Tradeoffs

Consideration	Voice Recognition	Dial Pulse Detection
Recognition /Detection	Language Dependent	Not Language Dependent
Accuracy	Speaker Dependent	Phone Network Dependent
User Friendly	Requires Discipline	Normal Dialing
Cost Per Channel	High	Moderate

Despite its lackluster image when compared with voice recognition, dial pulse detection has proven to be an effective technology that bridges the gap between past, present and future generations of telecommunication equipment for the following practical reasons:

- Dial pulse detection is universal not language dependent.

- The technology is mature and operational worldwide.

- Reliability and accuracy can reach 99% (although this may vary greatly between different products, suppliers and most notably telephone networks).

- Most people find dialing more natural when calling a computerized system. Speaking to a computer is less natural and sometimes embarrassing.

- No caller education or discipline is required with pulse detection.

- Pulse detection can be more cost effective and easier to integrate.

Integrating Dial Pulse Detection into CT Applications

When integrating dial pulse detection capabilities into an application, three issues need to be addressed:

- Distinguish between pulse and touchtone callers.

- Accommodate lack of "*" and "#" buttons on rotary dial phones.

- Accommodate slower dialing speeds.

Distinguishing Between Pulse and Touchtone Callers

In order to provide equal access to both DTMF and pulse callers, the developer must take into consideration the fact that these callers have different telephone instruments, with different capabilities, yet, they may look alike (push-button telephone that uses pulse dialing). Asking the caller to identify himself as a pulse caller, has proven to be impractical, and even confusing, as the typical caller has no perception of the signals his telephone instrument outputs.

Aerotel offers a concept which eliminates the need to ask the caller any questions, yet allows the CT system to recognize automatically, and definitely, the type of telephone used. According to Aerotel's method, all callers are asked by the CT system to dial a certain digit to proceed.

Example : "Thank you for calling ABC Company. To proceed, please dial zero."

If the caller dials from a touchtone telephone, Aerotel's dial pulse detection board will remain transparent, allowing the DTMF signal to pass through. If the caller dials from a pulse telephone Aerotel's board will recognize the '0' digit dialed, but then, output a DTMF digit '1' , thus signaling to the CT system that a pulse caller is on the line. Knowing this fact, the voice system can play a set of prompts different than the ones intended for DTMF callers (avoiding use of # and * signals, for instance) .

In addition to distinguishing between touchtone and pulse callers, the pulse dialed "0" is used to immediately analyze the telephone network characteristics and refine the pulse detection algorithms that will be used for the balance of that particular call.

Accommodating the Lack of "*" and "# " Buttons

Since Pulse callers can not produce the required # and * signals, an alternative has to be offered to those callers. Aerotel recommends implementing one of the 2 following methods.

- Time Out (preferred Method) - When the CT system identifies a pulse caller on the line no request for dialing "#" or "*" will be made. Instead, a time out will be used at the end of the expected input digits field.

- Special conversion - Aerotel offers a special option in its pulse detection products where a pre-designated combination of 2 pulse digits will output a DTMF "#" or "*" signal.

> Example :   Dialing '2 2' will output DTMF '#'
> Dialing '3 3' will output DTMF '*'

Please note that in applications where random numbers are used, any number with the above double digits in it will result in the output of the "#" or "*" signals. This is why this method can be used only in applications where the developer can control the input digit combinations.

Accommodating Slower Dialing Speeds

DTMF signals are precise and short. Normally a CT application calculates the string input time based on the DTMF signal duration. Pulse dials are significantly longer, especially on higher digits. Consequently the developer must allow at least one second per digit for pulse dialing signals.

Dial Pulse in the Order Status Application

Aerotel PTC 30D

The PTC 30D is an eight channel DSP based dial pulse detection board. One or more PTC 30Ds may be connected to loop start interface boards such as Dialogic's LSI/81SC or to E-1 interface boards such as the Auclab E1/ISDN Digital Network Interface. This is another example of standards based interoperability among CT component suppliers.

Training the Dial Pulse Detector

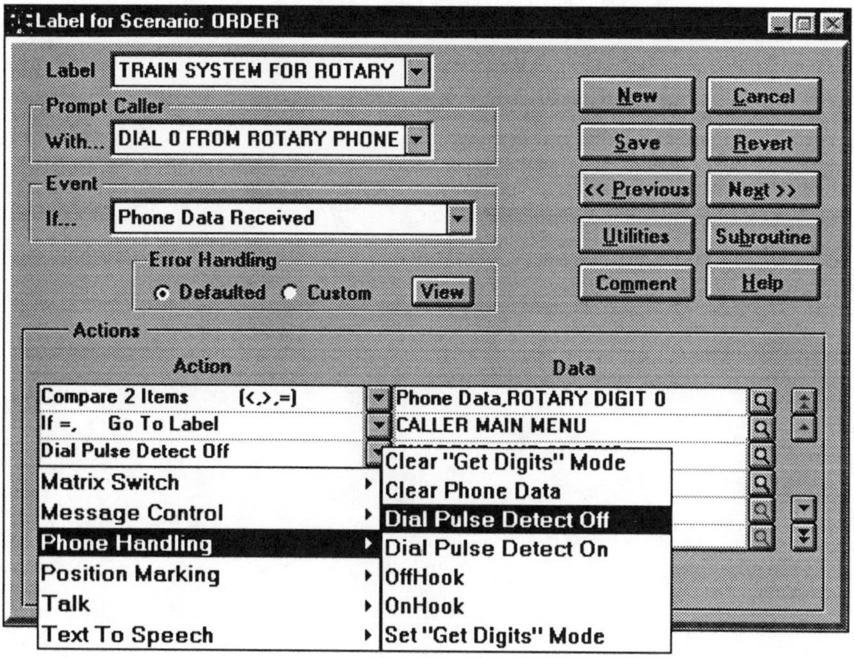

Figure 10-1

EASE Dial Pulse Detection

Figure 10-1 shows the order status application prompting the caller to dial zero to train the dial pulse detector.

After the CT application identifies that the caller is using a dial pulse phone and the detector is trained for that phone, the application branches to nodes of logic that accommodate the slower dialing speeds and lack of "*" and "#" buttons.

11

Automatic Speech Recognition

Automatic Speech Recognition

Automatic Speech Recognition (ASR) is how machines understand spoken input. The basic ASR process involves comparing a human "utterance" to a set of templates in a stored vocabulary to determine if there is a statistically valid match between the utterance and known vocabulary of the system.

Many ASR applications do not involve telephone networks. The voice typewriter is perhaps the best known. Also, man-machine dialogue remains one of the most popular ASR applications in film. One of the first characters to simulate ASR was HAL in the 1960's film "2001: A Space Odyssey". Today, "Star Trek", a popular science-fiction television program, continues exploring ASR.

Some ASR applications which do not involve telephone networks are:

- Hands-free factory-floor inspection stations

- Space Shuttle video camera controls

- Medical dictation systems

ASR Applications in Telecommunications

Speech recognition systems have been available for over twenty years, however, the technology has evolved slower than predicted. Only a few years ago, hardly any network-based speech recognition systems were in commercial use. This was because recognition technology over the telephone network was not sufficiently mature. Today, thousands of these systems are in operation and the number of new installations is growing rapidly.

Current speech recognizers offer cost-effective solutions for a variety of applications, such as:

- a **collect telephone call system** which determines whether to bill a collect call to a destination party, without the use of a human operator. The automated system integrates a telephone-network speech recognizer and a digitized speech recording in a pay telephone.

- an **ordering system** that allows a company's distributors to place orders over the telephone, without requiring human operators. The distributors answer a series of recorded questions and then a speech recognizer interprets the callers' spoken input.

- a **zip code information system** that provides zip codes to callers twenty-four hours a day, seven days a week. The telephone-network speech recognizer interprets a spoken address, confirms the address with the caller, and then provides the zip code.

Discrete Versus Continuous Word Recognizers

Speech recognizers may contain the following types of word recognition:

- **Discrete:** Recognition of a series of spoken words where more than 250 milliseconds of silence separates each word.

- **Connected:** Recognition of a series of spoken words where at least 50 but not more than 250 milliseconds of silence separate each word. Although confusing, many circles view **connected** as being synonymous with **continuous**.

- **Continuous:** Recognition of a series of spoken words where less than 50 milliseconds of silence separates each word.

Currently, a recognizer can have discrete, connected, or continuous word recognition. Also, discrete and connected word recognition may reside in the same recognizer. Plans are underway to provide discrete, connected and continuous word recognition in one recognizer.

Most applications of speech recognition use discrete word recognition, especially during menu selection or for "yes-no" type questions. When using discrete word recognizers, the recognition response time must be considered.

The recognition response time refers to the time it takes to recognize a word after the end of a word is spoken. In reality, the actual response time is longer because a certain amount of silence must occur before a recognizer declares the end of a word and begins recognition. Most applications can tolerate a recognition response time of up to 500 milliseconds for discrete words, as long as the caller is not entering, for example, a string of digits.

To illustrate the importance of recognition response time, imagine speaking to a speech recognition system. Picture the frustration a caller would experience waiting two seconds between speaking and the system responding with its next prompt.

When entering strings of digits spoken with more than 250 milliseconds of silence between each digit, the recognition response time becomes critical. It needs to be as short as possible, preferably 50 milliseconds. Discrete word recognition with a response time less than 100 milliseconds is referred to as connected word recognition.

A number of commercial speech recognizers are capable of recognition without explicit knowledge of word end points. These recognizers, referred to as continuous recognizers, tend to be somewhat more expensive than discrete word recognizers because they require much more intensive data comparison. Continuous word recognizers for telephone network applications are capable of recognizing the digits "zero" through "nine", and "oh". Currently, this is the only commercial speaker independent vocabulary available for network-based continuous-word recognition.

On the surface, one might argue that certain applications require a continuous digit recognition capability (where digits can be run together), simply because callers can speak naturally and input digits quickly. Because available continuous vocabularies do not include words like "fifteen" and "one-hundred-and-three", and because on-the-fly error analysis during continuous digit recognition is not currently possible, a fast discrete digit recognizer handles many applications more effectively than a continuous recognizer. It turns out that user experience is an important factor when deciding whether to choose discrete or continuous digit recognition. Keep in mind that an application using continuous recognition inherently uses discrete recognition also.

Speaker Dependent versus Speaker Independent Recognizers

Speech recognizers are either speaker dependent or speaker independent. With the state of today's recognition technology, both dependent and independent recognizers offer many benefits.

Speaker Dependent Recognizers

A speaker dependent recognizer is a recognizer that identifies spoken input after it has been trained for individual voices. All users must train speaker dependent recognizers before using them. This training involves:

- choosing a vocabulary

- each user training the recognizer to recognize his or her voice characteristics for the specified vocabulary

- each user repeating each vocabulary word several times in order to form a reference template

127

- each user training the recognizer in the appropriate environment(s)

After the user completes the training procedure, the recognizer will accommodate that user for the specified vocabulary.

Memory requirements are an issue with speaker dependent recognizers because vocabulary reference data is stored in the PC's memory. Factors affecting memory requirements include:

- the amount of memory per word

- the average number of trained vocabulary words per user

- the expected number of users

Recognition system performance reflects how well the users trained the system. Good training is critical if users expect the system to achieve high recognition accuracy. Although speaker dependent systems can support large vocabularies, it can require extensive and exhaustive training. Also, a system will experience performance degradation when a user trains the system in one environment and subsequently uses it in another.

Advantages of speaker dependent recognizers are:

- A user can change or add words to a vocabulary by simply training words on the recognizer.

- A dependent recognizer can support large vocabularies.

The major disadvantage of speaker dependent recognizers is that they limit the number of users.

Speaker Independent Recognizers

A speaker independent recognizer is a recognizer that identifies spoken input without users having to train it. The recognizer works for anyone's voice for specified vocabularies. Speaker independence is a critical requirement for telephone network applications because the identity of the caller is not known generally.

Speaker independent recognizers should provide high recognition accuracy under demanding network environmental conditions, regardless of sex, dialect, and other speaker characteristics. Important vocabulary words for most speaker independent applications are:

- "yes"

- "no"

- digits "zero" through "nine"

- "oh" (American English alternate for "zero")

Most commercial telephone network based recognizers offer these vocabularies as a minimum. Commercial technology is available that is accurate enough for virtually all applications requiring digit entry. Unlike

129

speaker dependent recognition over the telephone network, speaker independent vocabularies do not (and should not) require any type of adaptation after system installation.

Developers usually create speaker independent recognition vocabularies from large speech databases. What constitutes a good speech database for developing speaker independent vocabularies for use over the network? In general, the larger and more representative the database, the better it is. The public telephone network is a challenging environment for automatic speech recognizers. Factors that affect recognition performance are:

- Microphone variation

- Line conditions

- Signal band-limiting

- Gain variation

For the purpose of training the recognizer, experts agree that approximately 1,000 speakers represent adequately the telephone network and speaker variations. This assumes the vocabulary contains the digits "one" through "nine", "oh", and a few control words. Developers must ensure that when recording the data, speakers make calls from a wide range of geographical locations. The database must reflect a balance of

- male/female callers

- various dialects

- local/long distance phone calls

A shortage of data results in degraded accuracy, particularly for conditions not adequately represented in the database. Speaker independent vocabularies created from small databases can exhibit adequate accuracy over the telephone network, provided that the vocabulary size is small and the words are distinct from one another.

Advantages of speaker independent recognizers are:

- The recognizer does not require training by its users.

- A single set of templates can accommodate multiple users thereby minimizing PC memory requirements.

Recognition Accuracy

Commercially available speech recognizers exhibit a wide range of performance. To assess performance, we need to define and measure recognition accuracy. Recognition accuracy depends on:

- Type of system: speaker dependent vs. speaker independent

- Type of word recognizer: discrete vs. continuous

- Difficulty of the vocabulary

- Quality of the technology

- Users' environment

Recognition accuracy refers to the percentage of time the recognizer correctly classifies an utterance. Although the concepts discussed here apply to both discrete and continuous recognition, the focus is on discrete word recognition accuracy.

A recognizer can make three types of errors:

- Substitution error

- Rejection error

- Spurious response error

The most critical error is a substitution error. A substitution error occurs when the recognizer substitutes an incorrect word for the spoken word. For example, say the active vocabulary for a recognizer includes the digits, "one" through "nine". If the speaker says a "nine", and the recognizer identifies it as a "two", then the recognizer has substituted a "two" for a "nine". Substitution error rates must be less than two percent for user acceptance.

A rejection error occurs when the recognizer does not classify a spoken word, but rejects it. When rejection errors occur, the caller simply repeats the word until the recognizer identifies it. Rejection errors are not as critical as substitution errors. Rejection error rates should be less than three percent. Substitution and rejection error rates tend to decrease as the caller becomes more familiar with the system.

A spurious response error occurs when the recognizer classifies a sound or invalid word as a valid word. For example, the recognizer identifies the sound of dropping a telephone handset as a valid vocabulary word. Or, the caller says a word that is not in the recognizer's vocabulary, but the recognizer acknowledges it as a valid word.

In general, the larger the vocabulary, the more likely it is that spurious response errors will occur. Correctly classifying spurious input is a big challenge for today's speech recognition technology. For example, if a caller responds to a "yes/no" question by saying "What?", the recognizer should reject the response and re-prompt the caller. When designing an application, the developer should include spurious input handling to deal with situations such as those just mentioned.

It is difficult to quantitatively measure the spurious response error rate for a recognizer. To do so would require representing all possible sounds in a database for the recognizer. If such a database did exist, spurious response error rates of fifty percent could still occur. This means that when spurious sounds occurred, the recognizer would classify one-half of them as words in its vocabulary.

Testing for recognition accuracy generally ignores two types of rejection errors that are considered caller errors:

- Signal-to-noise error: When the recognizer rejects spoken input because the caller does not speak loud enough.

- Gap error: When spoken input begins before the recognizer starts listening.

Voice Stop

Certain speech recognizers do not allow callers to speak during a voice prompt. Voice stop allows the caller to interrupt a voice prompt by speaking. The voice stop feature does not however, allow the caller's words to be recognized during a prompt. It is useful for experienced callers that encounter long menu selections, because they can activate recognition by saying "Stop", for example, followed by the spoken menu selection.

Voice Cut-Through

Voice cut-through allows the caller's input to be recognized during a voice prompt. This feature is an extension of voice stop. The caller can interrupt the prompt and have his or her words recognized at the same time. As the application plays the outgoing message and voice cut-through is active, the voice prompt is reflected back to the recognizer input where it is effectively canceled through various signal processing techniques.

The advent of voice cut-through represents a major step toward the ultimate interface between humans and machines. People are accustomed already to using the touchtone keypad to interrupt prompts and enter information. Clearly, the ability to speak the information during the prompt is a significant advancement.

High Rejection

High rejection is the ability to discriminate with high accuracy whether a word is in a vocabulary. Since automatic speech recognizers are designed to find matches between utterances and the recognizer's vocabulary, they have a difficult time rejecting sounds or words that should be rejected. This is especially true for speaker independent recognizers when the vocabulary is somewhat difficult.

The implications of this depend on the application and the experience of the caller. For example, experienced callers are virtually unaffected because they rarely produce out-of-vocabulary sounds. In contrast, a first-time caller may, for example, attempt to clear their throat before speaking which can result in a recognition error.

A well known application that requires high rejection involves placing collect calls from pay phones. The system recognizes "yes", "no", and their synonyms, such as "yeah", "yep", and "nope". This application may involve interaction with a person not expecting to talk to a machine. Common responses to the prompt "Please say 'yes' to accept the call," may include "What?", "Am I talking to a computer?", "Hello", and so forth. Since the system is asking a person if they will pay for a long distance phone call, it is very important that the recognizer reject words and sounds not in its vocabulary.

For small vocabularies, it is relatively easy to achieve high rejection. As the vocabulary size increases, it becomes more difficult to maintain high rejection.

Word Spotting

Word spotting is identifying specific words spoken during natural unconstrained speech. Word spotting may also refer to the ability to ignore extraneous sounds during continuous digit recognition. In general, the term Word Spotting is used when the speaking mode is continuous.

Today, word spotting capabilities are limited because only small vocabularies can be accommodated. Some applications assume that a caller will respond to a prompt with one word from a small vocabulary. For example:

- The system prompts the caller with "What type of a call would you like to place?"

- The caller responds with either "collect" or "I want to place a collect call".

- The recognizer determines that the caller said "collect", either spoken in isolation or within a phrase.

Ideally, the recognizer can also determine that the caller said none or more than one of the vocabulary words. If this happens, the system re-prompts the caller. This of course requires high rejection, which the developer may or may not have designed into the recognizer.

In many applications, it is desirable to use word spotting when recognizing continuous digit strings. In general, the application needs to know the expected string length. This greatly simplifies the task of ignoring extraneous sounds during digit entry. Ideally, continuous digit recognition capabilities with high rejection for unconstrained string lengths will someday become available. However, it will be some time before applications reject the letter "A" instead of recognizing it as an "8".

Vocabulary Definition

A recognition vocabulary is the set of words a recognizer can recognize in any given application. For example, a common thirteen-word vocabulary includes "yes", "no", the digits "zero" through "nine", and "oh".

A recognition sub-vocabulary is a subset of a recognition vocabulary. For example, the words "yes" and "no" may comprise a sub-vocabulary of the above vocabulary. The digits "zero" through "nine" may comprise another sub-vocabulary.

A sub-vocabulary is the set of words that can be active at one time in an application program. It is the set of words for which the recognizer listens at a given point in an application.

Vocabularies range in size from one word to over 30,000 words depending on the nature of the technology and the application. Sub-vocabularies range in size too, but recognition accuracy requirements generally limit them.

Most telephone network applications require small vocabularies, say less than 50 words. Applications that involve dictation, do not involve telephone networks, and need quiet environments require large vocabularies.

Speaker Independent Recognition Vocabularies

Vocabulary requirements fall into one, or a combination, of the following categories:

- **Standard (Basic 16):** The digits "zero" through "nine", "oh", "yes", "no", "help", "cancel", and "stop".

- **Alphabet:** The letters "a" through "z".

- **Continuous:** The digits "zero" through "nine" and "oh".

- **Application Specific:** Checking, Savings, Mastercard, etc.

These categories apply to American English and other languages.

For most languages, the digit set in the standard category includes only ten vocabulary words. For example, the Spanish digits include "uno", "dos", "tres", "cuatro", "cinco", "seis", "siete", "ocho", "nueve", "cero".

Depending on the language, the digit set can contain more than one word for the same digit. For example, the British language uses "zero", "oh", and "nought" for the same digit. Japanese contains multiple words for the digits 4, 7, 9, and 0. Korean has multiple words for the digits 1, 2, and 0.

Languages

Standard (Basic 16) speaker independent, discrete vocabularies currently supported by Voice Control Systems, Inc. for use with Dialogic VR/160 ASR boards include:

Current ASR Languages

Afrikaans	Greek
Arabic (US)	Hebrew
Cantonese	Hindi (US)
Catalan	Hungarian
Danish	Italian
Dutch	Italian-Swiss
English-Australian	Japanese
English-British	Korean
English-Canadian	Mandarin (US)
English-Hong Kong	Polish (US)
English-Singapore	Portuguese-Brazil
English-South African	Portuguese-European
English-U.S.	Russian
Finnish	Spanish-South American
Flemish	Spanish-Castilian
French-Belgian	Spanish-North American
French-Canadian	Swedish
French-French	Tagalog (Filipino)
French-Swiss	Thai
German	Turkish
German-Swiss (High)	Vietnamese (US)

Vocabulary Reference Tables

Recognition vocabularies are represented in firmware in a variety of formats commonly called vocabulary reference tables. Perhaps the most important characteristic of a vocabulary reference table is memory storage efficiency. Customers typically ask recognition vendors about maximum vocabulary sizes, both the maximum number of vocabularies active at one time and the maximum total vocabulary size. The maximum number active at one time depends on:

- Memory

- Accuracy

- Response time

The type of recognition technology that the recognizer uses determines the memory requirements for each word in a vocabulary. On average, each word needs about one Kbyte of memory, but the range is quite large. Speaker dependent word representations require much less memory than speaker-independent words. This is because representing speaker independent words requires more statistical information.

From a practical point of view, accuracy and response time are important factors that determine maximum sub-vocabulary size. Increasing vocabulary size decreases accuracy, although word confusion influences accuracy significantly. Response time increases as the number of active words increase.

Operational Grammar: Using Sub-vocabularies

The most effective way to manage vocabularies in speech recognition applications is to incorporate operational grammars. An operational grammar (often referred to as syntax control) is a set of rules that determine which vocabulary words can be active at any given time.

For example, a caller could use voice recognition over the telephone network to make bank transactions. First, the CT system asks the caller to input his or her account number. The active vocabulary contains the digits zero through nine. The CT system then asks what type of transaction the caller wants to make. A possible active vocabulary might include the words transfer, balance, credit, and deposit. The basic concept is to only attempt to recognize words that need to be recognized.

Incorporating operational grammars into speech recognition applications results in many benefits:

- **Improved recognition accuracy**. Recognition accuracy can improve by an order of magnitude. The larger the vocabulary, the greater the chance that the recognizer will be confused.

- **More flexibility when selecting vocabulary words**. An application can use words that sound alike, as long as they are not active at the same time. Using the words "write" and "right" requires an operational grammar.

- **Quick response times**. The response time of any recognizer increases as the active vocabulary size increases. Operational grammars lead to small active vocabularies which result in quick response times.

- **Small vocabularies easier to remember**. Large active vocabularies are difficult to remember. In most cases, users find it difficult to remember a hundred words at a time. So, from a human factors standpoint, operational grammars are desirable.

Vocabulary Confusion

Recognition accuracy depends on the confusion of the vocabulary words trying to be recognized. It is well known, for example, that the letters of the alphabet are difficult to recognize with high accuracy. Because the success of any ASR application depends on recognition accuracy, it is beneficial to provide guidelines for selecting sub-vocabulary words.

In general, words with more than one syllable are easier to recognize than monosyllabic words. This is true because longer words usually contain more acoustic information than monosyllabic words.

Callers pronounce some words differently, particularly with speaker independent vocabularies.

Words that are distinguished from one another based on their beginnings (e.g., "no" and "oh") are easier to recognize than words distinguished from one another based on their endings (e.g., "seem" and "see"). People tend to say words in a manner that causes acoustic energy to decay near the end of the word. Usually, the beginning of a word can be automatically detected more reliably than the end of the word.

Custom Vocabularies

A major advantage of speaker dependent technology over speaker independent technology is vocabulary flexibility. This is why voice-dialing applications use speaker dependent technology. Voice-dialing allows a caller to say the name of the person you wish to call. In voice-dialing applications, individuals create their own personal speaker dependent vocabularies. However, ongoing demand exists for products that allow application developers to create their own speaker independent vocabularies. Also, they want to define and develop these speaker independent custom vocabularies without collecting large databases.

Speaker independent vocabularies created from small databases exhibit adequate accuracy over the telephone network if the vocabulary size is small and the words are distinct from one another. Commercial products are available that allow customers to develop their own speaker independent vocabularies. To develop a custom vocabulary, a developer needs to collect a database with approximately 50 to 100 speakers. However, testing the vocabulary requires more data and may end up being a larger task than developing the custom vocabulary.

The primary issue of concern when creating custom vocabularies is accuracy. Keep in mind that research laboratories with large, well designed databases will yield the most accurate vocabularies. When possible speech recognition experts should develop speaker independent vocabularies. The production of a vocabulary development system inherently results in something relatively simple when compared to the research laboratory facilities.

Microphone Issues

Words spoken into a microphone ultimately produce recognition vocabularies. A microphone is a transducer that converts instantaneous sound pressure waves into an analog electrical signal. For telephone network applications, the analog electrical signal may be transmitted over the telephone network after which it is converted into digital format before automatic speech recognition occurs. In some cases the audio signal is digitized in the network prior to the recognition site. For non-network applications, there is a direct analog audio path to the recognizer where the signal is then digitized.

Microphones exhibit a number of important response properties that are often overlooked and poorly understood, even by experts in the field of speech recognition research. Every speech recognizer utilizes a microphone. Recognition performance depends on microphone type and placement. It is important to understand basic microphone issues as they relate to speech recognition.

Telephone Network Recognition Environment

Discrimination of words over the telephone network is a challenging task, whose success depends largely on how well and how consistently speech signals can be analyzed and identified after having passed through a variety of different types and qualities of telephone microphones and transmission channels. Moreover, it is important to note that the success of any telephone network recognition system depends critically on adequate handling of the inherently wide variations in telephone signal quality that are introduced by varying microphone and line conditions. Telephone network recognizers have reliable performance because the algorithms, features, and parameters of each system are optimized specifically for the telephone network environment. Every process and parameter should be designed for robustness, i.e. for insensitivity to the wide range of speaker, microphone, and transmission characteristics that are typically encountered on the telephone network.

Figure 10-1 is a simplified block diagram of the telephone network environment as seen from the perspective of a speech recognizer.

Speech Recognition Over the Telephone Network

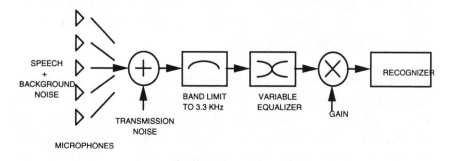

Figure 11-1

In this environment, a human produces speech in the presence of background noise. The noise can be moderate, such as quiet office noise, or intense, such as the noise typically encountered at an outdoor pay phone near a busy street. A telephone microphone with unknown acoustic properties converts the speech and background noise into an electrical audio signal. Before describing the rest of the recognition network environment, we consider microphone variability.

The major classes of microphone input media for telephones are:

- carbon button handsets

- electret handsets

- electret speaker phones

Each of these input devices possesses unique acoustic properties that have a significant effect on the way speech sounds to a listener at the other end of the network. For example, the frequency response of a carbon button handset microphone is quite different from that of an electret handset microphone.

Speaker phones produce signals that include a significant amount of echo because the acoustic energy of the speech signal at the microphone diaphragm is relatively low compared to the level of echo. Telephone network-based speech recognition systems must accommodate these and many other differences. In addition to the typical microphone responses, significant atypical characteristics may be encountered, e.g., from microphones in public pay telephones that have been abused.

The microphone output signal, which has been characterized as speech superimposed upon background noise, is transmitted over the telephone network where another component of noise, transmission noise, is added. This noise may be composed of high frequency tones, 60 Hz and other EMI, crosstalk, white noise, transmission echo, cracks and pops. The signal is also band limited from approximately 300 Hz to 3.3 kHz, which in itself makes the speech less intelligible. In addition, variable equalizers are often used to help line conditions meet required specifications. Finally, from the recognizer's perspective, an overall gain factor can span a range of 30 dB. For instance, it is not unusual for a human to have to call someone back because of a bad connection.

Alphanumeric Recognition Technology

Recognizing individual alphanumeric characters over the telephone is not an easy task. Even humans have trouble distinguishing the letters b, c, d, e, g, p, t, v, and z, especially over the telephone network. However, by application of a special recognition strategy, high alphanumeric string recognition accuracy can be achieved in most cases. By assuming that a spoken alphanumeric string is a member of a known, finite set of valid alphanumeric strings, a large number of invalid strings may be categorically eliminated as recognition hypotheses. In this way, individual character errors can be corrected to yield the most likely valid string.

147

Voice Control Systems has developed a new recognition technology accommodating vocabularies including the letters of the alphabet and the digits "zero" through "nine". These vocabularies have been developed for the purpose of automatically recognizing alphanumeric strings (i.e. strings comprising only digits, only letters, or a combination of both) spoken over the telephone network.

It has proven to yield remarkable performance improvement over individual character recognition, even for strings containing only two characters. As surprising as it may seem, longer strings are generally easier to recognize than shorter ones because more information is available to determine which string (from a finite set of reference strings) was spoken. The number of valid string candidates is critical since, given a fixed string length, as the number of candidates increases, the accuracy decreases, and the required computation increases.

The method for recognizing a spoken alphanumeric string involves assigning recognition distances between each spoken input and the corresponding letter or digit in the same position within each string represented in the database. Each recognition distance is a measure of the acoustic dissimilarity between a spoken input and a hypothetical character. For example, if an "A" is spoken, then the recognition distance for "A" is expected to be quite low. It is also likely that the distances for characters that sound similar to "A", such as "8", "H", "J", and "K", will be higher but also fairly low, and that distances for highly dissimilar characters such as "9", "Q", and "W" will be high.

The alphanumeric strings contained in the database are referred to as reference strings. After the caller says the first alphanumeric character, each reference string is assigned a distance value equal to the recognition distance between the spoken character and the first reference character of that string. After the caller says the second alphanumeric character, each reference string distance value is incremented by an amount equal to the distance between the second spoken character and the second reference character of that string. This process continues, accumulating distances for each reference string, until the caller says the last character. Then, the reference string with the lowest cumulative distance is the recognized string.

This recognition technique, called score-based recognition, can be used for discrete or continuous recognition. For continuous digit recognition, a database of valid digit strings must also be available.

The Disadvantages of ASR

With all this success, one might ask, "Why not install ASR in every application?" Well, good reasons exist not to:

- Speech recognition is not always as convenient or accurate as other forms of input to telephone voice processing systems. DTMF is often faster and more accurate for digit string input or for interrupting voice prompt messages for experienced users. However, in new applications speech recognition may be more useful for users.

- The general population has a greater awareness of how to use DTMF input. Most people are comfortable with DTMF and less comfortable with any new methods. They have to work to learn how to use ASR whether it is discrete or continuous. They must learn to wait for the beeps and they must learn to speak in a somewhat unnatural manner. However, advances in technology will rapidly overcome this limitation.

- ASR is more expensive for the hardware, software and product integration. The use of ASR contradicts the cost issues in cost sensitive markets like voice mail. However, ASR expands the range of possible applications where cost is not the critical determining factor.

- ASR is relatively difficult to implement from the system provider's and application programmer's perspective. However, that is because ASR is new technology. As the experience base grows and the product matures, implementation will become much easier. Also, aids, such as this book, help a system's implementation and the application programmer.

- ASR is a new technology with improvements and new features being rapidly released. As with most new technology, sometimes system hardware and software suppliers lag behind the innovations and changes. Features such as voice cut-through that significantly enhance performance of voice recognition will hasten the pace for supply.

ASR is not the perfect solution, but it is a valuable tool set to use when its capabilities match the requirements of a prospective voice processing application. An important point is that ASR is a tool set and not a single tool.

More Information on ASR

For more information on ASR, refer to <u>Speech Recognition - The Complete Practical Reference Guide</u> by Peter Foster and Dr. Thomas B. Schalk published by Flatiron Publishing, Inc. (formerly Telecom Library, Inc.).

ASR in the Order Status Application

Dialogic VR/160 with VCS Speech Recognition

The VR/160 ASR board offers four, eight, twelve or sixteen channels of speaker independent speech recognition in a single PC slot. The VR/160 board uses speech recognition algorithms from Voice Control Systems (VCS). VCS is the world's largest supplier of speaker independent recognizers.

Their extremely robust and accurate recognition vocabularies are based on an extensive database of over 100,000 speakers. The vocabularies are created specifically to withstand the marginal line conditions frequently found in demanding telephony environments. VCS technology is resident in more than 85,000 recognizers for telecommunications applications in 27 countries.

EASE Speech Recognition

Figure 11-2 shows speech recognition used as a touchtone replacement in the order status application. Instead of expecting an Event of 5 Touchtone Digits, the CT applications is looking for 5 Voice Recognition Digits.

This implementation uses basic 16 word, discrete speaker independent recognition.

EASE Speech Recognition

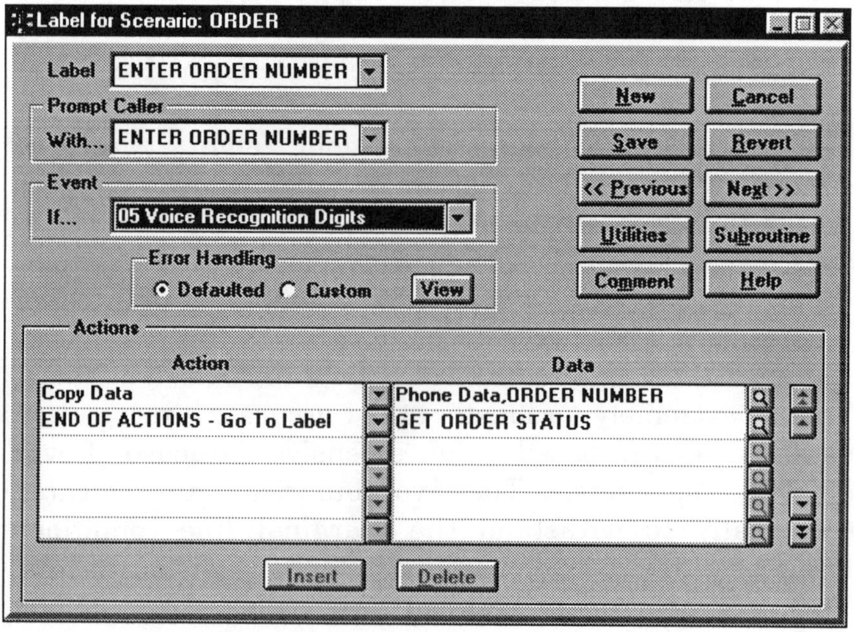

Figure 11-2

IV|
Computer Telephony System Output

12|
Digitized Speech

13|
Speech Concatenation

14|
Speech Synthesis

15|
Automated Fax Processing

16|
Call Transfer and Out-Dialing

17|
Serving the Hearing Impaired

12|
Digitized Speech

About Digital Audio

As recording engineers have known for years, it takes a little art, and more science, to capture sound and store it for later playback. You may have already tried to do some recording of your own and noticed that home recordings often don't sound as good as your favorite CD. Professional recordings typically sound clean and free of both distortion and background noise.

If you'll settle for fuzzy tones, distortions, random pops, thumps, buzzes, and hums, then just about any old microphone and recording technique will work. On the other hand, if you prefer digital recordings that are pleasing to the ear, then you'll want to acquaint yourself with some essential concepts. But before jumping into digital audio, it's important to have an understanding of the physics and perception of sound, and the process by which sound can get in and out of your computer.

Like a Pebble in a Pond

Sound consists of vibrations, either in the air or in some other medium. When a sound is created, waves of vibration spread out from a source, much the way waves spread out on the surface of a pond when you throw a pebble into it.

Waves have an up-and-down motion in pools of water. *Sound waves*, however, consist of variations in air pressure. The "crest" of each wave is a region where the air molecules are packed more closely together, and the "trough" is a region where the molecules are farther apart. This idea is illustrated in Figure 12-1.

Sound Waves Consist of Compressed (Crest) and Uncompressed (Trough) Air

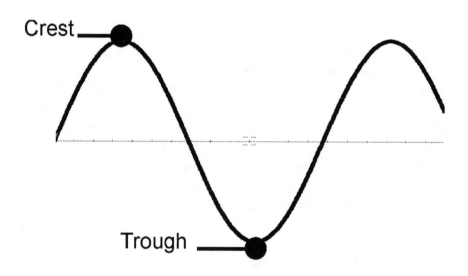

Figure 12-1

Waves are Busy, But don't Go Anywhere

In a pond, the waves travel outward, but the water itself doesn't go anywhere. All that happens is that the water molecules near the surface of the pond move up and down. You can see this if you float a dry leaf on the surface and then toss in a pebble. The leaf will not travel away from the point where the pebble entered the water; it will only bob up and down. In the same way, when sound waves travel outward from a sound source, the air molecules only press against one another while staying pretty much where they are.

Sound waves travel considerably faster than the waves in a pond: About 1,000 feet per second in air. And, if you've ever watched a marching band outdoors, you've probably noticed that you can see the cymbal players' cymbals crash together well before you hear them. That's because light travels a lot faster than sound. (About 8,900,000 times faster!)

Consider a loudspeaker. When it starts to make sounds, the speaker diaphragm is first pushed outward, compressing the air molecules. Then the diaphragm is pulled back, reducing the air pressure momentarily. As it is moved forward and backward again and again, sound is generated.

The key question is this: How quickly does the sound source move back and forth? If it vibrates rapidly, the peaks of the sound waves will be close together; if it vibrates more slowly, the peaks will be farther apart, as in Figure 12-2.

The Bass Fiddle has a Low Frequency and the Clarinet has a High Frequency

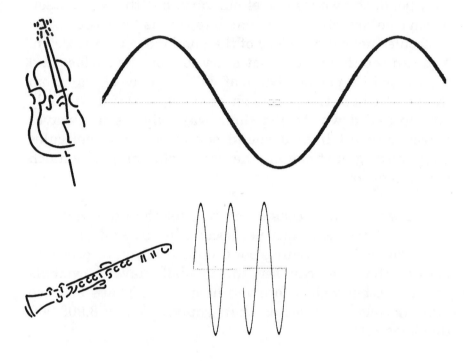

Figure 12-2

Our Ears Discern Subtleties

Human ears can very easily detect how closely spaced sound waves are when they arrive at our eardrums. When the sound source is vibrating rapidly, we say that the sound has a high *frequency*, because lots of waves reach our ears in a short period of time. When the vibrations are slow, the sound has a low frequency, because fewer waves strike our eardrums in the same amount of time. A clarinet, for example, produces higher frequency sounds, while a bass fiddle produces much lower frequency sounds.

Frequency is measured in cycles per second. The technical term for this measurement is *hertz* (abbreviated Hz). One cycle per second equals one hertz. Our ears can't detect sounds as low as 1 Hz—the low range of human hearing is about 20 Hz. At the upper end, children and adults with acute hearing can hear sounds as high as 20,000 Hz. (This is often abbreviated as 20 kHz - pronounced *kilohertz*)

The range of human hearing then, is usually considered to be about 20 Hz to 20 kHz. Many adults, however, have a reduced ability to hear high frequencies. They may not be able to perceive sounds above the 12 kHz to 15 kHz range.

Music

Frequency and musical pitch are closely related. Each musical note consists of vibrations at a specific frequency. For example, when you play a Middle C, you'll hear a sound whose frequency is almost exactly 261.653 Hz. The A above Middle C, which is often used as a tuning reference by orchestras, vibrates at 440 Hz. (Hence, it's called A-440.)

The musical scale sounds the way it does because the frequency doubles each time pitch rises by an octave. For example, the A directly below Middle C has a frequency of 220 Hz, and the next A below it vibrates at 110 Hz. Moving up the keyboard, the C above Middle C has a frequency of about 523.3 Hz, the next C has a frequency of

159

1,046.6 Hz. But if a clarinet, a bass fiddle, and a banjo each sound a note whose pitch is Middle C, how is it that our ears can instantly tell which instrument is playing?

Overtones

Consider the body of a banjo. When a note is played on a banjo, everything vibrates: the front of the banjo vibrates, the sides and back vibrate, the neck and fingerboard vibrate, and so on. The sound of a banjo, then, doesn't consist of a pure sound wave at a single frequency. Instead, each part tends to vibrate in a different way, and the sound produced by the instrument is a composite, or blend, of a number of different vibrations at different frequencies. If the sound of the banjo is recorded and examined on a computer screen, it won't look at all liked the simple sound waves shown in Figure 12-3. It will have a much more complex shape. Most sounds in the real world exhibit complexities of this kind.

Most Sounds in the Real World are Complex Waves

Figure 12-3

A mathematically pure tone is called a sine wave. Scientifically, it's possible to analyze a complex wave as being the sum of a number of different sine waves, each with its own frequency and amplitude (loudness). This is based on *Fourier's* (pronounced "FOOR-ee-aye") *Theorem*. The body of a banjo produces a complex wave containing a number of separate tones at different frequencies, all at the same time.

These are called overtones, and virtually all sounds in nature contain numerous overtones. One way our ears can instantly tell the sound of a banjo from the sound of a clarinet or bass fiddle is by noticing the relative loudness of the overtones at various frequencies.

Usually, the loudest of the overtones is the one vibrating at the nominal pitch of the note (440 Hz if the violinist is playing an A above Middle C). This frequency is called the fundamental.

With this understanding of what sound is, and what happens when it's propagated through the air, let's look at the process of capturing it on your computer.

Capturing Sound

A microphone is a form of *transducer*, a device that can take energy in one form and translate it into another form. Changes in air pressure arrive at a microphone, and are translated into changes in electrical voltage. During playback, a loudspeaker driven by electrical voltage transforms the voltage back into air pressure—so a loudspeaker is a transducer too.

Mounted inside a microphone is a small, thin, sensitive piece of material called the *diaphragm*. As sound waves strike the diaphragm, it vibrates at the same frequencies as the sound. The diaphragm, and its associated parts, translate these sound movements into fluctuating voltages.

The microphone needs to be built with precision so it can be as sensitive as possible to slight sound vibrations. Professional recording studios think nothing of spending $1,000 or more on a good microphone. You may not want to rely on the free microphone shipped with a sound card. Many of these microphones are for recording sounds to jazz up your computer operating system, but that's about it.

Analog to Digital Conversion

A microphone works much like a human ear. When sound strikes a microphone, the microphone produces an electrical wave that is analogous to the air pressure that produces the sound. Because of this correspondence between the air pressure and the voltage coming out of the microphone, the wave is termed an analog signal. "Analog" is used to characterize some common types of audio components, such as tape recorders, etc., to distinguish that equipment from the kind that incorporates digital circuitry.

Computers are designed to be operated in the digital realm, where everything is a series of on or off voltages, formatted as bits. To store sound on a computer we must convert the continuously changing analog audio signal into digital data. The circuitry that transforms data from analog to digital, and vice-versa is termed the *DAC* (digital to analog converter).

At regular intervals a DAC instantaneously freezes the audio signal voltage and holds it steady while another circuit selects the *binary* code that most closely represents the sampled voltage. The DAC outputs a number in binary format (ones and zeros) that represents the input signal at any given instant in time. See figure 12-4.

Analog to Digital Conversion

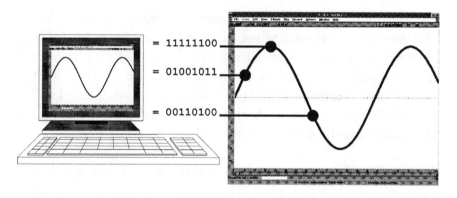

Figure 12-4

Digital-to-analog conversion (used in playback) is the exact opposite. In digital-to-analog conversion, the digital data is converted to a continuously changing series of voltage levels. The shape of this continuously changing stream of voltage levels approximates the shape of the original wave. This signal is then passed through a *low-pass filter*, which removes the digital "switching noise."

Once in digital form, the audio is extremely immune to degradation caused by system noise or defects in the storage or transmission medium (unlike old fashioned analog systems). The digitized audio signal is easily stored on a hard disk drive, where it can be kept indefinitely without loss.

Sampling Resolution

Sampling resolution refers to the number of discrete levels that are used in the analog-to-digital (and digital-to-analog) conversion processes.

Sampling resolution is measured in bits, which refer to the amount of memory required to store each individual sample. The number of different values that can be stored in a collection of bits is equal to 2 raised to the "bit'th" power. For example, an eight-bit sample is digitized to 2^8, or 256 different levels. A sixteen-bit sampling system, on the other hand, senses 2^{16}, or 65,536 different levels.

Obviously, the more bits of resolution that are used, the more closely the sampled signal will represent the analog signal, which has an essentially infinite resolution. However, higher resolutions require more storage and processing power, so some tradeoffs must be made. Eight bits is generally accepted as the lowest sampling resolution that can be used to obtain reasonable results, while 16-bit resolution is preferred for professional applications.

The resolution of a sampling system is almost always determined by the hardware. The minimum resolution permitted for multimedia hardware is eight bits. Compact disks and digital audio tapes use 16-bit samples, although many playback units only use 14 bits in their output circuitry. The telephone network typically uses 14 bits, but with an added twist called *compression*.

Compression

When the DAC outputs a binary number representing the input signal, this number must be stored in a convenient form for later retrieval. This conversion of a binary number for storage purposes is called encoding.

Audio encoding techniques can be broadly categorized into two classes: those for encoding analog waveforms as faithfully as possible, and those for minimizing (or compressing) the computer storage requirements. The two most common techniques used to encode an audio waveform are *pulse code modulation* (PCM) and *delta modulation* (DM).

Linear pulse code modulation (PCM) associates a particular binary number with every voltage level of the incoming analog signal. As the incoming signal increases, the binary number goes up in value proportionally. Similarly, as the analog voltage goes down, the binary number decreases. A multimedia 16 bit wave file is an example of linear PCM.

Non-linear PCM allows the computer to store fewer bits per sample by dropping the bits for signals that require less sensitivity. Non-linear PCM associates a particular binary number with a voltage range of the incoming analog signal. By carefully choosing the ranges, 14 bits of resolution can be packed into an 8 bit sample with minimal loss of fidelity. 8 bit u-law (pronounced "MEW-law", for the Greek letter u) is an example of a non-linear PCM encoding technique. 8 bit u-law is used to carry audio through the North American phone system.

Delta Modulation (DM) is a data compression method where only the difference between subsequent samples is stored. Since voice signals are relatively stable from sample to sample, the number of bits required to faithfully reproduce the signal can be reduced. One type of delta modulation is *Continuously Variable Slope Delta modulation* (CVSD), where a single bit is used to indicate whether the signal is increasing or decreasing. Another common type of DM is Adaptive Delta Pulse Code Modulation (ADPCM) where a constantly changing table of multiplier values allows the encoder to adapt to various types of signals. Dialogic 4 bit files are an example of ADPCM.

All you really need to know is that compression techniques exist and that their sole purpose is to reduce the number of bits required to store audio while simultaneously retaining as much fidelity as possible.

Frequency of Sampling

Every single number produced by the DAC fully represents the sound, but for a variety of reasons it is impossible to store every one of them in memory. This problem is handled by sampling the output from the DAC at a regular rate. This is called the *sampling rate*, and is an important factor in determining the quality of digital sound. The sample rate is also called the sampling frequency, because it too can be measured in hertz.

Because of the physics of electronic filtering, it is necessary only to sample a wave twice during each cycle to get an accurate representation. This principle holds true even for very complex sound waves. As we saw earlier, even the most complex wave is composed of the sum of sine curves at varying frequencies and amplitudes. Therefore, if you sample at a rate that is at least twice the highest frequency in your input signal, the content will be accurately captured.

If a signal is insufficiently sampled, new and unwanted frequencies are generated and added to the sampled sound. These are related to both the input frequencies and the sample frequency in such a way that they are virtually guaranteed to sound unpleasant. This is called *aliasing*.

Fortunately, aliasing is not a serious problem with most modern systems. Audio equipment incorporates special circuits, called *anti-aliasing* filters, that automatically restrict the bandwidth (frequency content) of the input signal based on the sample rate. For example, if the telephony card is commanded to sample a sound at 8 kHz, then the anti-aliasing filter is adjusted to reject frequencies over 4 kHz. A similar filter is used on the output when the sound is played back to smooth off the "rough edges" created when the analog data is digitized.

The sample rate used by a particular sampling system can usually be set through software, though the upper value is limited by the hardware. The Multimedia PC document issued by Microsoft specifies that the sound hardware be at least capable of sampling at 11.025 kHz and 22.050 kHz. Compact discs are recorded with a fixed sample rate of 44.1 kHz. Telephone quality boards, such as Dialogic boards, are capable of sampling at 6 kHz or 8 kHz, using compression techniques that result in 4 or 8 bits per sample.

At a relatively low sampling rate of 6 or 8 kHz (typical for telephone quality voice) far fewer code bits are produced each second than, for example, at the 44.1 kHz sampling rate used for CD's. For a two-channel 16 bit signal at a 44.1 kHz sampling rate, 11 million bytes are generated each minute. That's why you'll need at least an 800 MB hard disk to record an hour of compact disc quality music. On the other hand, a 60 second segment of compressed telephony audio, sampled at 8,000 samples per second, using 1 byte samples, translates into 500 thousand bytes of data, thereby requiring significantly less storage. Even with compression, though, you can see that most recordings take up quite a bit of disk space.

Digitized Speech in the Order Status Application

System vocabulary is determined by the application call flow. Figure 12-5 shows the Enter Order Number prompt from our order status example.

These prompts are collected in a Vocabulary Script to be used by the voice talent for the CT system. Recordings are usually made directly to disk now. However, a tape recording may be made first and then digitized.

EASE Order Entry Prompt

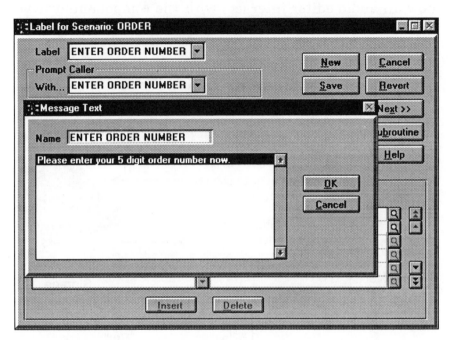

Figure 12-5

How Audio Editors Work

Now that you've stored the audio signal onto your computer hard disk, what can you do with it? Well, just as a word processor lets you manipulate the words and pictures that make up a document, an audio editor lets you edit sound in much the same way. You can trim off unwanted leading and trailing silence. You can select desired snippets of sound and discard the rest. You can make a sound "bold" by increasing its volume; make it "italic" by changing the pitch. With a graphical audio editor you can both see and hear the audio waveform while you work. Figure 12-6 provides a block diagram of how an audio editor interacts with the components you've seen so far.

How an Audio Editor Works

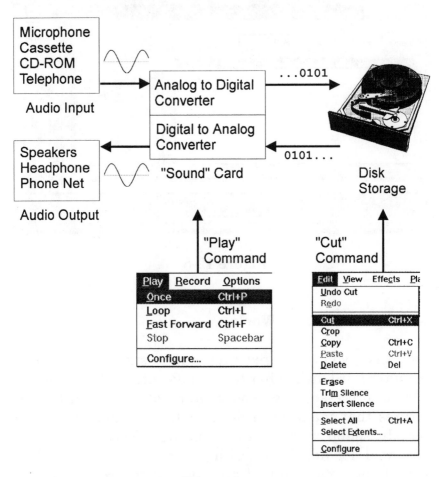

Figure 12-6

What is VFEdit?

VFEdit, from Voice Information Systems, Inc., is a graphical audio editor that lets you "see" the binary numbers that have been stored on your computer hard disk by your telephony or multimedia hardware.

VFEdit: Professional Prompt Editor

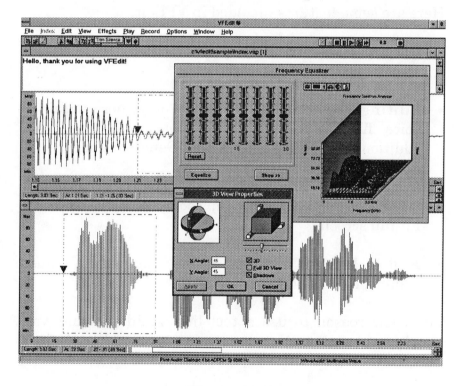

Figure 12-7

VFEdit knows how to translate these digitized and encoded audio signals into a picture that is meaningful to

171

you. Similarly, VFEdit knows how to take common editing commands such as "Play" or "Cut" and perform the requested operation on the hardware and the data. VFEdit provides a convenient user interface for recording new sounds and playing back audio already recorded. And VFEdit has been specifically designed to deal with the specialized hardware and audio formats used by telephony systems.

Here are some features included in VFEdit:

- **Play and Record:** Play and Record anywhere within a voice file.

- **Editing Tools:** Cut, Copy, Paste anywhere within a voice file; Copy To, Paste From and Mix From additional files. Erase, Crop, and Generate Tones.

- **Graphical Display:** View your voice file with full waveform display. Zoom In, Zoom Out and Zoom To Fit full screen.

- **Audio Effects:** Reduce noise, add echo, adjust volume, fade in and fade out.

The basic reason to use a tool like VFEdit is that you simply want your audio presentations to sound good—just like you want your written presentations to look good.

Professional Recording/Digitizing Services

It doesn't matter how much money you've invested developing a CT system, the way it sounds to callers greatly impacts your system's success. An organization's professionalism is reflected in the quality of the system's voice. For this reason, you may want to use a professional recording, digitizing and editing studio.

Some application generator vendors, such as Expert Systems, Inc. provide professionally recorded standard vocabulary with their basic software. If this is the case, you may want to consider contracting recording services through your vendor. This can assure high quality voice recording at a minimal cost since you will only pay for the incremental recording and digitizing that is required.

13|

Speech Concatenation

Speech Concatenation Fundamentals

Concatenation means linking together in series. *Speech concatenation* uses application logic to link together individually recorded and digitized words to produce a customized response in natural sounding language.

Speech concatenation uses a custom vocabulary that is usually stored on the hard disk drive and a standard vocabulary that is usually stored in RAM (Random Access Memory) for instantaneous access. These digitized words and phrases are linked together "on-the-fly" to speak numbers, currency amounts, dates, days, times and even to spell out words.

Custom Vocabulary

The following is a portion of the custom vocabulary from the order status example:

W#	Vocabulary
108	Thank you for calling the ABC Company.
109	Please enter your 5 digit account number now.
110	Your order number…
111	in the amount of…
112	was shipped on…

Standard Vocabulary for Speech Concatenation

W#	Word	W#	Word	W#	Word
1	Zero	37	February	73	A
2	Oh	38	March	74	B
3	One	39	April	75	C
4	Two	40	May	76	D
5	Three	41	June	77	E
6	Four	42	July	78	F
7	Five	43	August	79	G
8	Six	44	September	80	H
9	Seven	45	October	81	I
10	Eight	46	November	82	J
11	Nine	47	December	83	K
12	Ten	48	First	84	L
13	Eleven	49	Second	85	M
14	Twelve	50	Third	86	N
15	Thirteen	51	Fourth	87	O
16	Fourteen	52	Fifth	88	P
17	Fifteen	53	Sixth	89	Q
18	Sixteen	54	Seventh	90	R
19	Seventeen	55	Eighth	91	S
20	Eighteen	56	Ninth	92	T
21	Nineteen	57	Tenth	93	U
22	Twenty	58	Eleventh	94	V
23	Thirty	59	Twelfth	95	W
24	Forty	60	Thirteenth	96	X
25	Fifty	61	Fourteenth	97	Y
26	Sixty	62	Fifteenth	98	Z
27	Seventy	63	Sixteenth	99	Point
28	Eighty	64	Seventeenth	100	Beep Tone
29	Ninety	65	Eighteenth	101	Monday
30	Hundred	66	Nineteenth	102	Tuesday
31	Thousand	67	Twentieth	103	Wednesday
32	Million	68	Thirtieth	104	Thursday
33	Dollars	69	A.M.	105	Friday
34	Cents	70	P.M.	106	Saturday
35	And	71	Noon	107	Sunday
36	January	72	Midnight	1500	Silence

Figure 13-1

Sample Dialogue

If your order number is 12345, the order amount is $785.23, the ship date is 3/15/96, and computer telephony system logic is:

SHIPPED STATUS
"Your order number..."
speak Order Number as individual digits...
"in the amount of..."
speak Order Amount as dollars and cents...
"was shipped on..."
speak Ship Date as month, day, year.
Go to ORDER STATUS MENU.

Giving Data a Voice

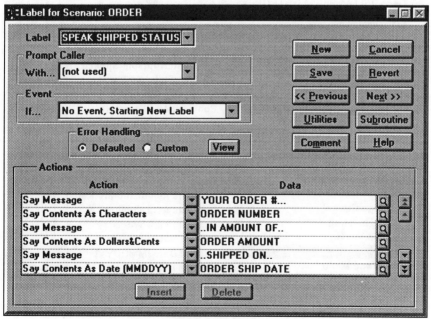

Figure 13-2

177

As illustrated in figure 13-2, the CT system would speak:

"Your order number one, two, three, four, five in the amount of seven hundred eighty five dollars and twenty three cents was shipped on March fifteenth, nineteen ninety six."

This is accomplished by the CT system linking together the phrases of the custom vocabulary and the words of the standard vocabulary "on-the-fly" based on application logic and preprogrammed talk functions shown in figure 13-3.

Talk Functions

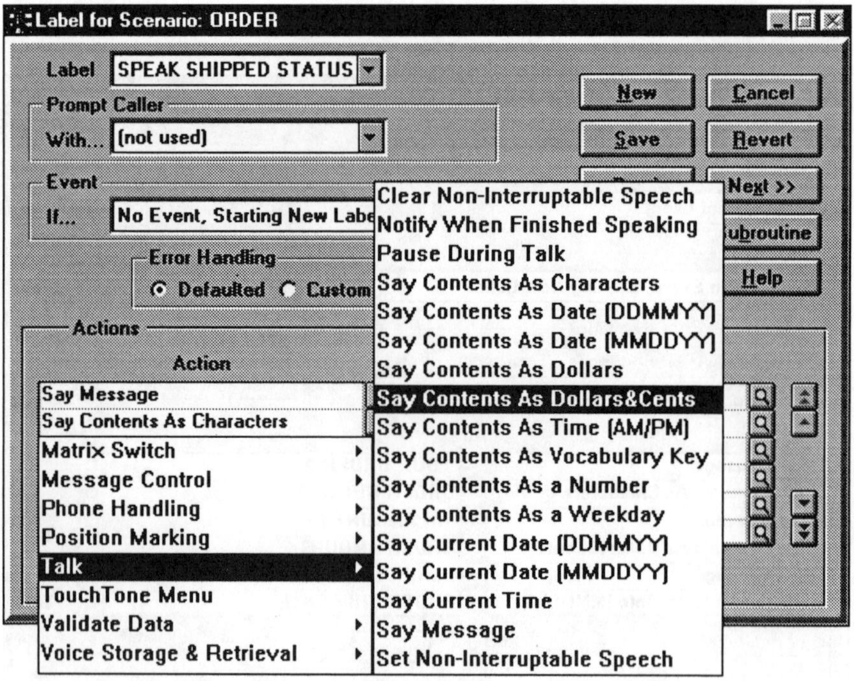

Figure 13-3

14|
Speech Synthesis

Text-to-Speech Synthesis

Text-to-speech synthesis (TTS) converts textual information into synthetic speech output. TTS is a knowledge-based artificial intelligence software technology that reads *ASCII* text and creates computer-generated artificial speech on demand.

TTS is made up of sophisticated rules that describe the way spoken language is created. It contains rules for predicting pronunciation from spelling, rules for deciphering text that is not phonetically pronounceable (Dr. vs. Dr., $12.34 vs. 1243%), rules for assigning sentence and word-level stress and intonation, and rules for generating the sounds that make up a particular language.

Text-to-Speech Versus Digitized Speech

Text-to-speech is used in applications to deliver information that digitized speech or speech concatenation can not reasonably deliver.

TTS's major role in CT systems is that it gives callers access to large or frequently updated databases. It is generally not a substitute for recorded human voice. If the vocabulary is small enough to record, then text-to-speech may not be the best delivery medium. But, if the CT system needs to deliver information which can not be recorded and digitized reasonably, then text-to-speech is the best way to do it.

Text-to-speech is fundamentally different from digitized human voice. In some ways, text-to-speech is more like a human operator than like the recorded voice in a CT system.

Few application developers would expect to put a CT system on line without first listening to all the voice messages in the script. In fact, most developers use professional voice talent to record each message, then digitize and edit the final recordings before releasing the system. Text-to-speech, on the other hand, has to take unedited text and do the same sort of job, but automatically, and in one "take."

Like a human operator, text-to-speech is expected to communicate information it has never seen before by pronouncing it correctly and understandably on the first try.

In a properly designed application using good technology, TTS can do this job about as well as the average human operator, and even better in some cases. Text-to-speech can be "taught" words that the average untrained person would not know. TTS only has to be told how to say things once, has a reliable (if somewhat synthetic) accent, and never takes a lunch break.

The Process of Text-To-Speech Synthesis

The TTS process is extremely complex, but can be summarized in eight steps. For reference, a *phoneme* is the smallest normally significant unit of sound in a particular language. In English, the 'p' sound in 'pit' and the 'sp' sound in 'spin' are phonemes. Also, *prosody* means intonation or the natural tonal ups and downs used when speaking.

1. **Text Input:** Text data made up of words, numbers, and other characters comprise the initial input for the TTS process. In a computer telephony context, this ASCII text could be in the form of a file such as a prompt for caller input or an e-mail message that has been stripped of header information, a variable such as an account balance or order status, or even a memory buffer containing the response from a host computer transaction.

2. **Text Normalizer:** The text normalizer converts everything in the text stream to letters. For example, in English '1996' becomes 'nineteen ninety six' and 'Dr.' becomes 'doctor' or 'drive' as determined by context. Normalization rules are different for different languages, and sometimes even for different countries that speak the same language.

3. **Exception Dictionary:** Some words, especially in English, are not pronounced in accordance with the basic rules for pronunciation. The English word 'of', for example, is the only case where a final 'f' is pronounced 'v'. To say these words correctly, a

phonemic transcription of their exact pronunciation is stored in the exception dictionary.

4. **Letter to Phoneme Rules:** Letter to phoneme rules convert normal spelling into phoneme transcriptions which are a more exact representation of their pronunciation. For example, 'cat' is represented as 'k ae t'. Each language has its own set of phonemes, prosody rules and phonetic rules.

5. **Prosody Rules:** Prosody rules create the intonation pattern or rhythm of the sentences. All computer synthesized speech sounds somewhat robotic because a computer must follow rigid instructions, unlike human speakers. Prosody rules make the intonation of the synthesized speech as pleasant and human-like as possible.

6. **Phonetic Rules:** The fine tuning of pronunciation takes place in the phonetic rules. For example, the 't' in 'toy' is subtly different than the 't' in 'cat'. Phonetic rules process these differences in an effort to make synthesized speech as clear as possible.

7. **Voice Generation:** The voice generation module converts phonemes into digital voice parameters. This is where pre-selected instructions modify the default voice to determine the character and even sex of the final voice.

8. **Speech Output:** The final step processes the digital information through a PC circuit board containing one or more digital signal processor (DSP). The DSP takes digital information and creates audible sound. This circuit board is connected to a voice processing board

which provides the telephone interface for text-to-speech synthesis.

Intelligibility

The best way to measure whether text-to-speech is being used successfully in an application is to measure whether or not the information is adequately understood by the caller. Users often assume that the main factor affecting this is the intelligibility ("understandable-ness") of the particular text-to-speech system being used.

When hearing a new text-to-speech system however, many people's first response is "Why doesn't it sound like a person?" It is useful to compare listening to text-to-speech to listening to someone with a Texas accent (unless you're from Texas, in which case, substitute a New York accent). It usually takes a few sentences to get the hang of the new accent, but after a few seconds, you can easily understand what is being said.

Designing Text-To-Speech into a CT Application

All text-to-speech systems have some kind of accent. After a sentence or two, most people find TTS as easy to understand as a human operator with a regional accent. Given this, the single most important factor affecting the success of an application is actually the design of the user interface.

In designing a text-to-speech application, it helps to apply some of the tricks human operators use to get their message across. These include allowing the caller to say "what?" ("To hear the information again, press one."), repeating the information more slowly, and offering to spell the individual words if the caller is trying to write the information down.

Also, remember that text-to-speech gives the caller access to much more complicated information than a digitized voice system typically does, so make it as easy as possible on the listener by not giving too much information at once. Where possible though, use full sentences. Like a person, text-to-speech sounds better when the words it is reading make sense.

Very often, developers new to text-to-speech reach first for the parameter resets which affect the sound of the voice (see Voice Generation above), thinking that they will be able to make it more intelligible by adjusting the pitch or speaking rate. After a few days of playing around they come up with an "improved" voice.

They play it for their customer, and are surprised by a negative reaction. What has happened is that they have not actually improved the voice, but merely traveled down the learning curve and gotten used to the text-to-speech accent.

Try the default voice first! Some people do actually prefer faster or slower speech, but other voice resets are usually best left alone. If the developer feels that they must change what the voice sounds like, get an outside opinion before it is demonstrated to a customer. If a developer is nervous about adding a synthetic voice to an application, try introducing it with digitized voice saying, "Now the computer will read you the information from our database."

User Interface Tips

Mixing digitized voice and text-to-speech in the same sentence is not recommended. The result is as difficult to understand as a sentence spoken by two different people. However, it is a good idea to use digitized voice wherever possible.

Generally, digitized voice should be used for caller input prompts and small vocabulary information. Text-to-speech should be used wherever digitized voice is not practical.

Some information is just hard to understand, no matter who reads it. If Bob Smith is verifying his own name, chances are good that he will recognize it perfectly, whether it is spoken by a person or by the computer.

If a delivery person is given "Chicken Lips Software, Inc." as his next pick-up destination, his reaction, whether he gets the information from a computer or directly from his boss, will be "What? He can't have said 'chicken lips,' chickens don't have lips!" In a digitized voice system the application developer picks words that are easy to understand, but text-to-speech and human operators do not have that option.

In any application where unusual information is being delivered, the developer should offer the caller the same options a human operator would. These may include one or all of the following: repeat, repeat slowly or with pauses between the words, spell-back for verification, or clear-spell ('c' as in Charlie) when the caller is trying to write down unfamiliar information.

Reading Names and Addresses

TTS will give a "correct" default pronunciation for names and addresses about as often as a human operator will, sometimes more often, given that human operators are sometimes quite far from the caller's region. This job can be difficult however.

People come to the United States from all over the world, bringing their names with them. People tend to make up new spellings and/or pronunciations for their names, depending on how long they have lived in this country, or even just based on personal choice.

Unless you know someone personally, you can not be sure that you know how to pronounce his name. A good example is Dialogic Corporation's John Alfieri. He may pronounce it "Al-feer-y," but someone else in his own family may very well say "Al-fee-air-y." Neither one sounds exactly like the original Italian.

The best approach to delivering large name and address databases with text-to-speech is to offer the same user interface options described above. For reading unfamiliar addresses, the "clear- spell" option is especially good, since it tells the text-to-speech to read the letters slowly, giving an example for letters which are hard to tell apart over the telephone (c, b, d, g, etc.).

No matter how correct the pronunciation, "Szyperski Street" will be impossible for a caller to write down if he has never been there. The "clear-spell" option ensures successful delivery of the information.

Changing Pronunciations

The Exception Dictionary (described in item 3 in The Process of Text-to-Speech Synthesis section of this chapter) is useful for names or application-specific abbreviations to which the application developer would like to assign pronunciations different from the TTS default pronunciation. The application developer may need to consult the linguistic staff of the particular TTS supplier to implement large or complex Exception Dictionaries.

Languages Supported by Text-to-Speech

The following languages are currently support by BeSTspeech TTS from Berkeley Speech Technologies:

- English

- French

- Spanish

- German

- Italian

- Japanese

- Dutch

- Russian

TTS in the Order Status Application

Dialogic TTS

The TTS/20, TTS/40 and TTS/80 boards offer simultaneous text-to-speech processing on two, four or eight channels respectively in a single PC slot.

The Dialogic TTS boards are based on BeSTspeech TTS methodology developed by Berkeley Speech Technologies, Inc., and complemented by proven Dialogic firmware technology.

EASE Text-to-Speech

Figure 14-1 illustrates an alternative to digitized speech concatenation when "Giving Data a Voice" in the order status application.

Notice that the entire order status is given using TTS instead of mixing TTS with digitized voice.

First the file containing "Your order number" is processed with TTS, then the variable containing the Order Number is processed with TTS, and so on until all information is conveyed.

When the TTS logic node is complete, the CT system will return to digitized speech mode.

EASE Text-to-Speech

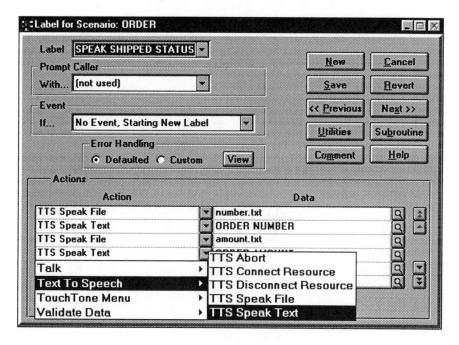

Figure 14-1

15|

Automated Fax Processing

Fax Standards

The beginning of *facsimile* technology is traceable to 1843, when the first successful fax device was patented. Commercial fax service began in France, in 1865. Over the 1930s and 1940s, fax evolved into the form we recognize today.

As fax became more widely used, it soon became obvious that standards would be needed to enable different fax machines to communicate with each other. In 1966, the *Electronic Industries Association* (EIA) proclaimed the first fax standard: EIA Standard *RS-328*, Message Facsimile Equipment for Operation on Switched Voice Facilities Using Data Communication Equipment.

This *Group 1* standard as it later became known, made possible the more generalized business use of fax. Although Group 1 provided compatibility between fax units outside North America, those within still could not communicate with other manufacturers' units or with Group 1 machines. Transmission was analog, typically it took between four to six minutes to transmit a page, and resolution was very poor.

U.S. manufacturers continued making improvements in resolution and speed, touting the "three-minute fax".

However, the major manufacturers, Xerox and Graphic Sciences, still used different modulation schemes--FM and AM--so, again, there were no standards. Then, in 1978, the *CCITT* came out with its *Group 2* recommendation, which was unanimously adopted by all companies. Fax had now achieved worldwide compatibility and this, in turn, led to a more generalized use of fax machines by business and government, leading to a lowering in the price of these units.

When the *Group 3* standard made its appearance in 1980, fax started well on its way to becoming the everyday tool it is now. This digital fax standard opened the door to reliable high-speed transmission over ordinary telephone lines. Coupled with the drop in the price of modems--an essential component of fax machines--the Group 3 standard made possible today's reasonably priced, familiar desk top unit.

The advantages of Group 3 are many; however, the one that quickly comes to mind is its flexibility, which has stimulated competition among manufacturers by allowing them to offer different features on their machines and still conform to the standard. The improvement in resolution has also been a factor. The standard resolution of 203 lines per inch horizontally and 98 lines per inch vertically produces very acceptable copy for most purposes. The optional fine vertical resolution of 196 lines per inch improves the readability of smaller text or complex graphic material.

Group 3 fax machines are faster. After an initial 15-second handshake that is not repeated, they can send an average page of text in 30 seconds or less. Memory storage features can reduce broadcast time even more. The new machines also offer simplicity of operation, truly universal compatibility, and work over regular analog telephone lines, adapting themselves to the performance characteristics of a line by varying transmission speed downward if the situation requires it.

The future of fax is evolving toward higher transmission speeds and added features, such as color. *Group 4* is a standard written to implement some of these features into future fax devices. However, Group 3 machines today do most of the things envisioned for Group 4 devices, and are compatible with the more than 35 million machines across the world, whereas Group 4 devices are not.

The Fax Call

With small variations, a modern Group 3 fax call is made up of five discreet stages as defined by the *ITU-T T.30* specification (see figure 15-1):

- **Phase A:** Call Establishment

- **Phase B:** Pre-message Procedure

- **Phase C:** In-message Procedure and Message Transmission

- **Phase D:** Post-message Procedure

- **Phase E:** Call Release

The Fax Call

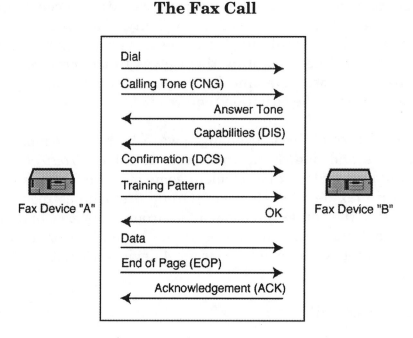

Figure 15-1

Phase A: Call Establishment

Phase A takes place when the transmitting and receiving units connect over the telephone line, and use a handshake procedure to recognize one another as fax devices. The calling fax device usually sends a medium pitch tone (1100 Hz) known as the *Auto Fax Tone* or *CalliNG Tone* (CNG). The CNG is on for 0.5 seconds, off for 3 seconds, and is repeated every 3.5 seconds for about 45 seconds after a number is dialed. Think of the CNG as a beacon identifying the calling device as a fax machine. As a side note, this is the tone that phone/fax/modem switches listen for to identify an income call as a fax call that should be routed to a fax machine.

When the called fax device answers, it responds with a wailing tone (2100Hz) to signal that it is ready to receive a transmission. This tone is known as the CallED Station Identification (CED). Anyone who has dialed a fax machine by mistake is very familiar with this sound.

Once the handshake is established, both devices move on to the next phase.

Phase B: Pre-message Procedure

In Phase B the called fax device identifies itself, describing its capabilities in a burst of digital information packed in frames conforming to the *High-Level Data-Link Control* (HDLC) standard. Always present is a Digital Identification Signal (DIS) frame describing the device's standard features. Two optional frames may also be present: a Non-Standard Facilities (NSF) frame which informs the calling device about any vendor-specific features in the called device and a *Called Subscriber Identification* (CSI or CSID) which contains and ID number - usually the telephone number of the called fax device. At this point, the called fax device may optionally transmit some analog signals to determine if the calling fax device is an older Group 1 or Group 2 fax device incapable of recognizing digital signals.

Assuming that both fax devices are Group 3, the calling device responds with a Digital Command Signal (DCS) frame informing the called device how to receive the fax. This information includes modem speed, image width, image encoding and page length. The calling fax device may also send a Transmitter Subscriber Information

(TSI) frame containing its phone number and a Non-Standard facilities Setup (NSS) command responding to an NSF frame.

At this point, the sending device activates its modem which, depending on line quality and capabilities of the devices, may use either *V.27* data rate to send at 2,400 or 4,800 bits per second (bps), *V.29* to transmit at 7,200 or 9,600 bps, or the new *V.17* for 7,200, 9,600, 12,000 or 14,400 bps. A series of signals known as a "training sequence" is sent to let the receiver adjust to line conditions, followed by a *Training Check Frame* (TCF). If the receiver successfully receives the TCF, it uses a V.21 modem signal to send a Confirmation to Receive (CFR) frame; otherwise it sends a Failure-to-Train (FTT) and the sender replies with a new DCS frame requesting a lower transmission rate.

Phase C: In-message Procedure and Message Transmission

Phase C is the fax transmission portion of the operation. This step consists of two parts-C1 and C2-which take place simultaneously. Phase C1 deals with synchronization, line monitoring, and problem detection. Phase C2 is devoted to data transmission.

Since a receiver may range from a slow thermal printing unit needing a minimum amount of time to advance the paper, to a computer capable of receiving the data stream as fast as it can be transmitted, the data is paced according to the receiver's processing capabilities. If both devices support *Error Correction Mode* (ECM), and ECM procedure encapsulates data within HDLC frames, providing the receiver with the capability to check for and request the retransmission of garbled data.

Phase D: Post-message Procedure

Once a page has been transmitted, phase D begins. Both the sender and receiver revert to using V.21 modem signals as during phase B. If the sender has further pages to transmit, it sends a frame called the *Multi-Page Signal* (MPS) and the receiver answers with a Message Confirmation Frame (MCF) and phase C begins all over again for the following page. After the last page is sent, the sender transmits either an End Of Message (EOM) frame to indicate there is nothing further to send, or an End Of Procedure (EOP) frame to show it is ready to end the call, and waits for confirmation from the receiver.

Phase E: Call Release

Once the call is done, phase E, the call release part, begins. The side that transmitted last sends a Disconnect (DCN) frame and hangs up without awaiting a response.

Computer-Based Fax (CBF)

GammaLink originated PC fax technology when, in 1985, it released its first PC-to-fax hardware and software products. Originally, fax boards were viewed just as a way for standalone PCs to emulate fax machines, and applications were limited to the transmission of ASCII files. Although fax transmission has always been used for highly visual data, poor quality reproduction was a fact of life until the advent of CBF.

CBF boards, like the ones manufactured by GammaLink, Dialogic and PureData, enable users to receive faxes, print them, store them on a hard disk, as well as to create them using text and graphics files, and then transmit them. Depending on system configuration, these boards can act as fax servers for vast computer networks, sending, receiving, and routing faxes for entire multinational companies.

One of the major advantages of CBF is in the quality of reproduction at the receiving end. Text and graphics that are run through a fax scanner are inevitably degraded in sharpness. It does not matter whether the reproduction medium at the other end is thermal paper or a laser printer, it is the scanning process at the transmitting end that degrades the image. Text and graphics produced on a PC, however, are almost invariably superior. Also, it is possible to preview received faxes on the computer display before printing or discarding them, an increasingly useful feature as fax comes to be used more frequently for purposes of every kind.

Another major advantage is in the capability to broadcast large numbers of faxes. Often, it may be necessary to do a mass broadcasting of faxes to clients, suppliers, etc. With CBF systems, this becomes a simple matter: a list is created, and the computer is instructed when to send out the faxes to the people and companies on the list.

This last feature is useful if the user prefers to transmit at a time when phone rates are at their lowest. Besides this automation feature, other benefits lie in the ability to track fax expenses as closely as any other business expense and the customization of CBF for different purposes, enabling the user to find novel applications for faxes as selling tools, marketing purposes and, in the case of fax-on-demand applications, to make information available to staff and customers at remote sites 24 hours a day.

One of the most advanced applications of CBF is host interactive fax. This exciting new technology allows on-line data to be merged with template forms such as invoices or account statements, and faxed to callers "on-the-fly". Customers can check the status of their order or shipment; distribution locations can receive up to the minute inventory levels from the manufacturing plant; customers can receive an interim account statement - all without talking to an operator. The latest information is sent to the caller's fax machine on the appropriate form. Even the cover sheet can be personalized by extracting customer contact fields from a database.

The Fax Image

The fax image is made up of dots. The dots cover the page and create letters, numbers, and pictures. Standard resolution is 204 dots per inch (dpi) in the horizontal direction and 98 dpi vertically. Fine mode doubles the dots per inch in the vertical direction making the resolution 204 x 192 dpi.

The more dpi, the clearer the image. However, increasing the resolution increases the file size that must be transmitted across telephone lines. Fine mode images take approximately twice as long to send as standard mode images.

Compression

When digital processing was developed for fax technology, it became practical to use encoding techniques to remove redundancy from the page being scanned, and restore the data at the receiving end. This helped reach the sought-after results of error reduction and shorter transmit times.

In a scan line a white picture element, or pixel, is likely to be followed by a long string of the same, before reaching a black pixel. In typed or printed material, these white strings (or runs) generally continue across the entire page.

A white or black pixel represents 1 bit of information. The early fax machines would transmit 1728 bits of information for every white line across a page. Now, instead of sending all 1728 bits to represent a white line across a page, *Modified Huffman* (MH) compression produces a 9-bit code word representing that white run, thus compressing this information 192 times. The result is that only 92 binary codes are needed for white runs of 0 to 1728 pixels, and another 92 for the black ones. The shorter binary codes are assigned to the longest, and most commonly occurring runs, in this way saving even more transmit time. The receiving fax units then decode this information, reproducing the original run.

A typical Group 3 fax image can have a total of 1,973,376 pixels-1728 from side to side and 1,143 from top to bottom. Were this image sent uncompressed at 9,600 bps, under ideal circumstances, it would take almost three-and-a-half minutes to transmit. This is not practical and it can get very costly, in terms of telephone time and personnel, to take over ten minutes to fax a three-page document.

Today, Group 3 fax machines use two compression technologies. One of these, the Modified Huffman code, is an encoding scheme that takes advantage of the similarities between pixels on the same line but not between pixels in succeeding lines. It uses a combination of run-length and static (not adaptive) Huffman encoding. In this process, the sending scanner searches for transitions from black to white (or vice versa) and reports the number of pixels since the previous transition.

To prevent blotching in the event of an error, the line must always start with a "white" run length (which can be zero if the first pixel is black). The code for a run of black pixels is not the same as for an equal number of white ones because they are not as statistically likely since runs of black pixels tend to be shorter than runs of white ones in most printed documents.

If the run is longer than 63 pixels, a special "makeup code" is prefixed to the code word. Each makeup code adds a multiple of 64 pixels to the run length. One makeup code plus one normal run-length code is enough for a run of up to 2623 pixels long (wider than a normal A4-page fax transmission).

Modified Huffman compresses only one scan line at a time; it looks at each scan line as if it were the only one. Each line is viewed by it as a separate, unique event, without referencing it in any way to previous scan lines. Because of this single-line operating characteristic, modified Huffman is referred to as a "one-dimensional compression" encoding technique.

The second encoding scheme, MR, or *Modified READ* (Relative Element Address Differentiation), uses the previous line as a reference, since most information on a page has a high degree of vertical correlation; in other words, an image, whether it is a letter or an illustration, has a continuity up and down, as well as from side to side, which can be used as a reference. This allows modified READ to work with only the differences--the variable increments or rates of change--between one line and the next.

This results in about 35 percent higher compression than is possible with Modified Huffman. Modified READ's operation may seem a complex concept to grasp, but it really isn't. Imagine for example, a white sheet of paper with a black circle in the middle. The sheet can be effectively compressed because, after the first compression, there is no change in the following lines. The modified READ algorithm just repeats the line.

When modified READ starts scanning the circle, black run lengths begin and, as it proceeds downward, the circle scan gets increasingly larger. However, since modified READ uses previous lines as a reference, instead of counting all the black pixels to encode them as the

Modified Huffman technique would, the modified READ algorithm only notes the rate of change between lines.

Therefore, the difference between Modified Huffman and Modified READ is, essentially, that the latter uses a "knowledge" of previous lines to reference its vertical compression. Because Modified READ works vertically as well as horizontally, it is called a "two-dimensional" compression encoding technique.

Two-dimensional encoding, even with all the necessary safeguards against errors, can reduce transmission time by as much as 50 percent. Actual sending time depends on the number of coded bits per line and modem speed.

The newest compression technique is call Modified Modified READ (MMR). MMR is a two-dimensional scheme for Group 4 fax that is being incorporated into Group 3 because of its error handling capabilities called Error Correction Mode. ECM allows the receiving fax device to check for garbled data and request a re-transmission.

As a cautionary note, in certain situations compression techniques can result in increased transmission time - achieving the opposite of the intended result. This typically occurs in images with a multitude of black and white changes both vertically and horizontally that require the transmission of many additional compression code words. This adverse effect is most noticeable with desktop publishing output that uses computer generated gray-scale.

Effective Transmission Speed

A traditional fax machine is a mechanical device. It must reset its scanner, and advance the page as it prints each scan line it receives. Today's machines generally have a 10 millisecond (ms) scanning line time requirement.

When fax devices negotiate during the handshake process before transmission, they exchange information on scanning capabilities. If the receiving machine's rate is 20 ms per line and the transmitting machine sends data at a faster rate, it will add *fill bits* (also called *zero fill*). These extra bits pad out the amount of send time, giving the remote machine the additional time it needs to reset prior to receiving the next scan line.

The amount of fill bits added by the sending machine is determined by the receiving machine's capability. If a machine that sends at high speeds transmits to a slower machine, some of the time saving benefits of the encoded data and higher modem speed may be lost if a lot of fill must be introduced.

The rate at which a machine can receive is not related to the rate at which it can send. Although computer based fax (CBF) devices generally do not require fill bits, if it is sending to a traditional fax machine, it must tailor its sending mode to that device's capabilities.

Getting Hard Copy

When the receiver modem decodes the received analog fax signal it regenerates the digital signal sent by the fax transmitter, the MH/MR/MMR block then expands this fax data to black-white pixel information for printing.

There are two ways to convert a fax into hard copy: through a thermal printer, or a regular printer (the latter covering everything from pin to the preferred laser printer formats).

Each inch of a thermal printer's print head is equipped with 203 wires touching the temperature sensitive recording paper. Heat is generated in a small high-resistance spot on each wire when high current for black marking is passed through it. To mark a black spot, the wire heats from non-marking temperature (white) to marking temperature (black) and back, in milliseconds.

About the only advantage of thermal paper is that, since it comes in rolls, a thermal printout has the advantage of adapting itself to the length of the copy being transmitted: if the person on the other end is sending a spreadsheet, the printout will be the same size as the original. With sheet-fed printers, such as a laser printer, the copy is truncated over as many separate sheets as necessary to download the information.

In general, thermal printouts offer less definition. It is inconvenient to handle due to its tendency to curl; and if the copy is to be preserved, it must be photocopied because thermal paper printouts tend to fade over time. Then, most important of all, there is the matter of cost: thermal paper printouts are between five to six times more expensive than those made on plain paper.

Plain paper fax machines using laser printer technology are becoming increasingly popular, particularly in the U.S.

File Formats: TIFF and PCX

When a fax machine sends or receives a fax, it goes from paper to paper, being converted to an image as described above. When a computer sends or receives a fax, it stores the image as a file or series of files. Three of the common file formats for storing fax images are *TIFF* (Tagged Image File Format), *PCX* (PC Paintbrush) and *DCX* (multi-page variation of PCX). ASCII files can also be faxed but these are not considered images.

TIFF, PCX, and DCX formats are widely used by other computer programs such as drawing, graphics, and scanning packages. The fact that standard formats are used gives the user a lot of flexibility in manipulating fax images, combining them with other images, and adding them to word processor documents.

A TIFF file of the same image typically takes up about half the space of a PCX file. If you plan on deploying a system that will either contain or receive a lot of files, you may want to remember that TIFF files will require about half the disk space of systems using PCX and DCX.

The average image size will generally range from 20 to 40K per page for TIFF files and 40 to 80K per page for PCX and DCX files. The amount of space required is dependent primarily on the resolution of the page and the composition of the image. Pages containing text will usually be much smaller than pages containing pictures or graphics with large areas of gray.

International Certification

The Post Telephone and Telegraph (PTT) administrations, usually controlled by their governments, provide telephone and telecommunications services in most foreign countries where these services are not privately owned.

It is not a simple thing to obtain approval from the PTTs to sell and use telecommunication equipment of any kind in their countries. The world is far from being one in the field of telecommunications.

Meeting international requirements typically means providing hardware and software modifications to the product, unique to each country, and then going through an extremely rigorous approval process that can average between six to nine months per country. The approval process requires products to meet both safety regulations and compatibility requirements of each country's telephone systems.

For More Information on Fax Processing

For more information on fax processing, refer to Computer Based Fax Processing - The Complete Guide To Designing And Building Fax Applications by Maury Kaufman published by Flatiron Publishing, Inc.

Interactive Fax in the Sample Application

Dialogic VFX/40E-SC

The VFX40E-SC board offers four ports of enhanced voice/call processing and 14,400 bps (V.17) fax services in a single PC slot. This dual capability gives caller unrestricted use of voice and fax processing in a single call. The fax channels may be used as resources dedicated to particular voice lines or they may be shared by multiple voice lines through the SC bus interface. The VFX/40E-SC's on-board loop start interface is approved in over 25 countries.

The Interactive Fax Process

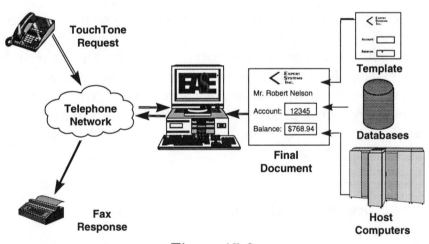

Figure 15-2

EASE Interactive Fax

The VFX supports the TIFF/F file format and ASCII-to-fax conversion on the fly. This feature allows real-time data to be inserted in to a fax template and faxed to a caller on demand. See figure 15-2.

Figure 15-3 shows the development process of inserting the variables for order number, order amount and order shipped date into the order shipped fax template.

EASE Interactive Fax

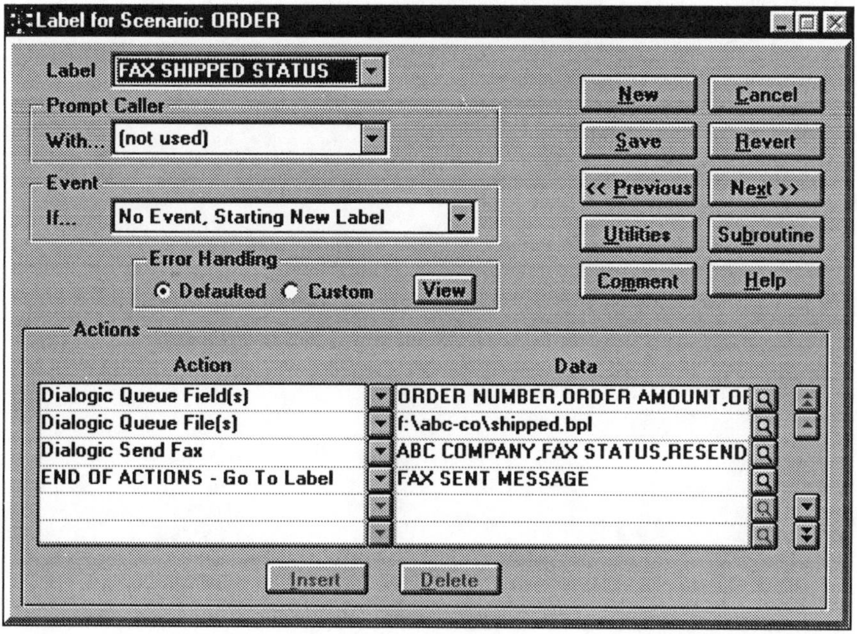

Figure 15-3

At this point you may want to go back to the order status demonstration line instructions in Chapter 3 and request a fax confirmation of your fictitious order. This is a truly exciting new technology. And it is real!

In fact if your organization and fax number are in Expert Systems' current contact database, your fax confirmation will be personalized with the name and address of the primary contact in the database. This is a nice way of proving that real-time data is being merged with a fax template on-the-fly.

16|
Call Transfer and Out-Dialing

The Call Transfer Process

In many cases computer telephony systems are implemented to only handle routine information entry and inquiry calls. This means that non-routine and problem calls must be transferred to an operator, agent, or representative for handling with the human touch.

Call transfer requires a switch! The switch can be Centrex service provided by the phone company central office, an on-premise PBX, or digital timeslot switching performed within the CT system. In any case, a switch is necessary.

You are probably familiar with the call transfer process using a PBX. It is basically the same when performed by a CT system regardless of whether it is an analog or digital switch connection. The steps are as follows:

- Issue a hookswitch flash to put the caller on hold. This also serves as a request for service from the switch

- Dial tone is presented by the switch

- Dial the destination number of the transfer

At this point in the process you could either:

1. Hang up and assume that the transfer will go through; or,

2. Stay on the line and listen for various signals such as ringing or busy to help assure that the transfer goes through.

In the latter case you could wait for a voice to answer, converse with that person regarding the call, and then hang up, which takes the caller off hold and connects them with the destination line.

However, if you heard a busy signal or ring/no answer, you could do another hookswitch flash, take the caller off hold, and choose another course of action.

The first type of transfer described above is called blind call transfer and the second is called intelligent call transfer.

• **Blind Call Transfer:** In a blind call transfer, the CT system relinquishes control and final disposition of the call to a telephone switch or to another computer telephony system.

• **Intelligent Call Transfer:** With intelligent call transfer, the CT system maintains application control over the entire call transfer process applying "intelligence" as required to assure that the caller receives the best available disposition at that time in a particular environment.

Blind Call Transfer

Blind Not Unintelligent

Application intelligence can be applied to the transfer process prior to initiating a blind call transfer.

For example, a caller may request to be transferred to technical support, however, the CT system may have used Automatic Number Identification or an account number entered by the caller to determine that the caller is not covered by a current support plan. In this situation, the CT system could override the caller's request and transfer them to sales instead of technical support.

Blind call transfer simply means that the CT system initiates the transfer and then relinquishes control of the call completion process. The CT system initiates the blind call transfer by dialing an extension and then simply hanging up. Typical devices to which the CT system may relinquish control of the call progress include:

- **PBX with Hunt Group Capability:** A hunt group is a designated series of phone lines organized in such a way that if the first line is busy, the next line is hunted and so on until a free line is found. The hunt may start with the first physical line and hunt sequentially, or it may start randomly and hunt clockwise or counter-clockwise. The point is to find an available agent who can take the call transfer, but this determination is made by the PBX not the CT system that initiates the transfer.

- **Automatic Call Distributor (ACDs):** ACDs are devices used by companies that handle a large volume of calls, such as airlines or telemarketing firms. An ACD answers calls and directs them to specific agents who answer them. The heart of an ACD is a software program that determines how to route calls based on pre-established criteria. For example, an incoming call might be directed to the first available agent; or calls from certain parts of the country can be routed to specific agents (the origin of a call can be determined using ANI). Most ACDs provide a monitoring function so that a supervisor can determine such things as whether calls are being answered promptly or how many calls an individual agent handles in a given period of time. ACD software will also provide management reports containing statistics on calls received (date, time, line number), average waiting time before calls are answered, number of calls, abandoned calls, etc. A CT system is not an ACD in most cases, and will not typically provide the functions and statistics mentioned above.

- **Automated Attendant:** Automated attendants are CT systems that answer incoming calls and play a recording. Most commonly, the recording is a greeting ("Thank you for calling the ABC Company") followed by a menu of choices (For Sales, press 1; for Customer Service, press 2; for Accounting, press 3; or, stay on the line for an operator"). Callers can respond by pressing the appropriate touchtone digit on the keypad of their telephone. Callers who have rotary telephones are normally routed to an operator when they do not respond to the menu prompt with a touchtone within a given period of time. Automated attendants can be used in place of a receptionist, or to answer the telephone after hours or on weekends. They can also be configured to answer calls when the receptionist is busy on another call. Most CT systems can be programmed to handle the automated attendant function.

These devices are usually backed up by voice messaging systems or subsystems for situations when no operator or agent is available.

Intelligent Call Transfer

Call Progress Analysis

When you stay on the line while transferring a call, you use your ear and brain to do some sophisticated information collection and analysis. And based on this analysis, you make branching decisions.

If you hear a busy signal or ring/no answer, you may stop and try a different number. If you hear ringing that stops after a couple of cycles and then you hear a voice, you would proceed with the transfer. However, your decisions are only as good as your knowledge of the various signals and what they mean. This knowledge of signals and branching logic must be imparted to the CT system if it is going to supervise call transfers.

It is important to point out that *call progress analysis* is not an exact science. The identification of the various signals is probabilistic and often ambiguous in the time frame allowed for analysis. In addition, the branching logic must accommodate all possible outcomes of the analysis...and there are more outcomes than you think!

Recognizing Signals

Voice boards typically use two types of call progress analysis:

- **Cadence Based:** For most call transfers within a local PBX, it is sufficient to recognize only four outcomes: no ring back, busy, ring/no answer and connect. This can be accomplished by teaching the CT system to recognize the timing of the rhythmic patterns produced by the busy signal and the ringing signal. ring/no answer is defined as a maximum elapsed time, and Connect is identified as a break in the ringing cadence. The major drawback to cadence based call progress analysis is the analysis duration. It usually takes several seconds to identify the presence of a cadence. Another drawback is that once the CT system has identified that ringing cadence has started, it will interpret any break in cadence, such as automatic call forwarding, as a connect.

- **Frequency and Cadence Based:** For more sophisticated PBX call transfers and out-dials that go through the public telephone network, an analysis of various frequencies on the line, in addition to the cadences, can provide more information beyond that available from simply examining cadences. Frequency analysis can also provide results with a much shorter analysis period.

Certain call progress parameters can be pre-defined based on known cadences and frequencies while others must be "learned" by the CT system operating in the actual production environment. This can be a complicated process that involves playing samples for the CT system in a "learn" mode. Each CT system must be trained for each particular switch because different switches use different parameters.

Possible Outcomes of Call Progress Analysis

For instructional purposes, the following is a list of possible outcomes of an out-dial or call transfer through the public telephone network (see figure 16-1). The frequencies and durations described below do not necessarily apply to a local PBX; however, the CT system can be trained to identify a wide range of specific frequencies and associate them with call progress events.

The general order of the possible outcomes is: failures, incompletes, and connects.

- **No Dial Tone:** No dial tone is presented by the switch or central office during a specified period of time. Since North American dial tone is a combination of 350 and 440 Hz, the frequency analyzer would typically be set to look for 400 Hz with a deviation of +/- 125 Hz before proceeding to dial. If dial tone is presented, the actual frequencies detected would be reported by the voice board to the CT system. This would allow the CT system to determine whether it was local, international, or a special dial tone.

- **Special Information Tones (SIT):** You have probably heard successive tones of increasing frequency followed by a digitized voice message saying "The number you have reached is not in service". The three tones signal prior to the message is one of four special information tones.

SIT	Description	1st Tone	2nd Tone	3rd Tone
NC	No Circuit Found	985.2	1428.5	1776.7
IC	Operator Intercept	913.8	1370.6	1776.7
VC	Vacant Circuit	985.2	1370.6	1776.7
RO	Reorder -System Busy	913.8	1428.5	1776.7

- **No Ring Back:** Ring back is the sound you hear when you're calling someone else's phone. Ring back cadence typically is 2 seconds on and then 4 seconds off. The signal is generated by a device at your phone company's central office and may bear no relationship to the sound the phone at the other end is emitting or not emitting. Ring back can be detected by cadence or frequency. Typical, application control in a no ring back outcome is to re-dial the number.

- **Busy:** A "slow" busy signal indicates that the phone at the other end is busy or off-hook. The slow busy signal is typically in the 500 Hz range and is applied in a cadence of 0.5 seconds on, then 0.5 seconds off, 60 times per minute. The "fast" busy signal is applied in a cadence of 120 times per minute when the telephone network is congested with too many calls. With fast busy, the phone at the other end may or may not be busy; but your call did not progress far enough to find out.

- **Ring/No Answer:** As mentioned earlier, ring back cadence is 2 seconds on and then 4 seconds off, in a 6 second duty cycle. Duty cycles are not usually counted. ring/no answer is typically defined as a maximum elapsed time such as 30 seconds after ring back cadences is identified.

- **Connect:** In it simplest form, a connect can be defined as a break in ring back cadence which indicates that the phone has been answered. By examining the *answer size* (the length of time between the break in ringing cadence and a subsequent period of silence), you can make an "educated guess" about who or what answered. A short answer size (under 1 second) could be presumed to be a residence because the typical greeting is simply "Hello" or "Smith residence". A medium answer size (1.5 to 3 seconds) is generally a business with a greeting such as "Good Morning. ABC Company." Finally, a long answer size (over 3 seconds) could indicate an answering machine with a greeting like, "Sorry we're not in. Please leave a message at the tone and we will get back to you." Or, a long answer size could indicate a fax machine or modem. Of course, this method is not absolutely accurate because greeting messages can vary greatly, and a pause in the middle of the greeting could be falsely detected as the end of the greeting, thus yielding an incorrect answer size.

- **Connect to Fax/Modem:** If you employ frequency as well as cadence analysis to the connect, the CT system can identify a fax machine by the continuous 2100 Hz CallED Station Identifier (CED) tone emitted by a fax device when it answers. Most modems can also be identified by their continuous 2150 Hz auto answer carrier tone.

- **Connect to Human Voice:** The frequency analyzer, such as Dialogic's Positive Voice Detection, can be set to detect audio signals that have the characteristics of a human voice (fluctuating frequencies in the 300 to 3000 Hz range), and thus identify that a call has been answered. This is a very precise method for identifying when a connect occurs. However, if you want to try to determine whether the human voice is live or recorded, you must do additional frequency analysis.

Call Progress Analysis Outcomes

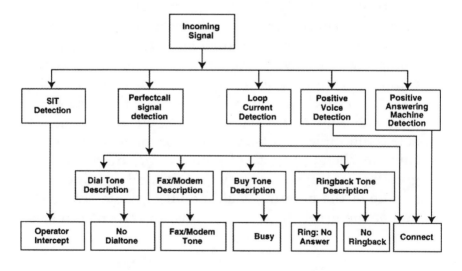

Figure 16-1

- **Connect to Answering Machine:** The frequency analyzer, such as Dialogic's Positive Answering Machine Detection, can be instructed to look for tape "hiss" and generic beep tones to increase the reliability of answering machine detection. This provides greater accuracy than answer size analysis; but is not always reliable because of the wide range of answering devices.

Call Transfer in the Sample Application

Design Tips

Common sense is important in implementing good automated call transfer. When a CT system puts a caller on hold and dials an extension using call progress analysis, the caller does not hear the destination party say "hello" because they are still on hold. If the CT system automatically completes the transfer after using the "hello" as a connect detection, then both parties will typically be sitting in silence on the line.

This can be avoided by programming the CT system to mimic a transfer like a human operator does it. After the CT system detects a connect, it should say something like "I have a call for you" and then complete the transfer. This lets the destination party know that they must say "hello" again.

Of course, instead of just saying "I have a call for you", the CT system could provide more valuable information. For example, it could preface a transfer with "Credit hold transfer. Account number XXXXXXX." This provides much of the value associated with screen pop interfaces.

EASE Intelligent Call Transfer

Figure 16-2 shows the node of logic called Transfer To Operator.

EASE Intelligent Call Transfer

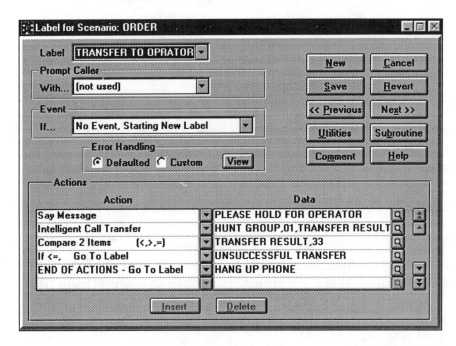

Figure 16-2

At the bottom of the screen, you can see that the caller will hear the message, "Please hold for an operator." The

Intelligent Call Transfer action is used to transfer the call to the appropriate hunt group using transfer model 01. Model 01 was constructed when the CT system was in learn mode and contains the specific intelligent call transfer parameters that were set up for the particular switch.

If the CT application receives a transfer result of "33", the transfer is considered unsuccessful and the application branches to a node of logic called Unsuccessful Transfer. If the transfer result is successful, the CT application completes the transfer process by simply hanging up the phone.

17|
Serving the Hearing Impaired

The Americans with Disabilities Act of 1990

There are two reasons why you should consider adding *Telecommunications Device for the Deaf* (TDD) access to your CT applications and services. First, it's the right thing to do. Second, in some cases, it's the law.

The Americans with Disabilities Act of 1990 (ADA) requires that all public entities with automated telephone services provide equal access to the deaf, speech or hearing impaired. The ADA also requires that telecommunications service firms provide relay services, i.e., operators to relay calls between TDD and voice telephones. A three year grace period was allowed for, so ADA requirements are law as of 1993.

TDD access will appear on Request For Proposals (RFPs) for any government agency and on many from larger corporations.

What is a TDD?

TDD is a more modern term for the teletypewriter or TTY. A TDD at a minimum has a keyboard and a small character display, and sometimes uses an acoustic coupler to connect to the phone line. A TDD uses in-band modem tones to communicate over standard PSTN lines.

The modern TDD provides memory to save conversations, has larger displays and even a printer. They sell for about $200 to $500 and are available at many consumer electronic retailers. Single line TDD interface boards are also available for the PC. These are convenient for the computer literate, but are not as portable.

How does TDD Work?

TDDs use a half duplex, frequency shift key (FSK) modulation scheme at rates of 45.5 baud domestically, and 50 baud internationally. Characters are encoded into five bit *Baudot code*. Unlike data modems, TDDs do not use a continuous carrier. Instead, a sequence of five FSK tones are generated each time a character is sent. Most modern TDD devices are also capable of higher speed communications, utilizing ASCII and standard modem protocols, from 100 baud on up.

Since Baudot code has only five bits, only 32 separate and distinct characters are possible from this code (2x2x2x2x2=32). By having one character called Letters (marked LTR) which means that all subsequent characters are alphabetic characters and one character called Figures (marked FIGS) meaning that all subsequent characters are numerals, punctuation marks or special characters, the Baudot character set can represent 52 (2x26) printing characters. "Space", "Carriage Return", "Line Feed" and "blank" mean the same in either LTRS or FIGS.

Lack of Standards

There is no official standard for TDD communications. The Telecommunications Industry Association (formerly EIA) attempted to standardize, but failed because the TDD manufacturers (only two existed at the time) would not agree. More recently, a Draft Recommendation was issued by the CCITT (now the ITU) for a "V.txp" standard for text phones used by the disabled.

What about DTMF?

If an application can't demodulate the incoming TDD modem, then how does the caller control it? With DTMF. It *is* acceptable to assume that a TDD caller also has a DTMF phone attached to the line.

TDD Store and Forward

TDD Store and Forward is possible on many of today's voice products. TDD modem signal, being in-band, half duplex and non-continuous carrier, are easily recorded, stored and played back as a voice file. Use of 64 kbps, No Automatic Gain Control, record mode is recommended. Care should be taken to disable or lengthen any silence time outs.

The application developer will be required to obtain a TDD device - used to "record" prompts. Incorporating TDD into an existing voice application will require all of the voice prompt scripts to be typed-in using the TDD.

A TDD device may also be required at the end user application site. For example, a business or government may be required to retrieve and respond to a customer request left as a TDD message.

TDD store and forward will enable message, TDDtext (like audiotext) and Interactive TDD Response applications.

Please be advised that the above procedure is covered by patents (US patent 5,121,421 and 5,253,285) issued to Curtis C. Alheim and licensed through DiRad Technologies, Inc. Curt Alheim has also developed a specialized workstation (US patent 5,450,470) that insures the proper translation of a voice script into the "printed" equivalent script for TDD transmission and use. For more information, contact DiRad Technologies in Albany, NY at 518-438-6000.

TDD Modulation

TDD Modulation - taking a text string and automatically generating the FSK tones - would make application development quicker and easier. Technically, this feature is not very difficult to implement in a DSP, although it is complicated somewhat by the lack of standards. Current voice boards typically do not support this.

TDD Demodulation

TDD Demodulation - receiving FSK tones and decoding into a text string - is technically one level more difficult than TDD modulations. Few applications may require this feature. Current voice boards do not typically support this.

Do I Need a Separate Line?

Will my application require a separate phone line dedicated to TDD? Publishing a separate phone number and dedicating a line may be positive and reassuring to the TDD caller. However, it may not be practical for a smaller (few lines total) application.

An alternative is to request the TDD caller to press a DTMF key to switch to TDD mode. Instead of publishing a separate phone number for TDD callers, publish an instruction for them to dial the main number and then "Press # for TDD access." Remember that the caller may not be able to hear the voice greeting, so this would not be a good place for such an instruction!

For More Information on TDD

The Telecommunications Industry Association offers their latest draft of a TDD specification, dated March, 1988. Their efforts to finalize this specification have since been abandoned. Nevertheless, there are some very applicable sections. TIA @ 202-457-4936

The CCITT has issued "V.txp, a draft recommendation for TDD standard.

The Civil Rights Division of the US Department of Justice runs an automated information line, to assist the public in understanding the act and implementing the

V|
Accessing Data

18|

Accessing IBM Mainframe Data

IBM 3270 Product Series

IBM 3270 is not a specific product, but instead denotes a terminal oriented family of devices designed to work in clusters. The key components for emulation purposes are the 3178/3278 display terminal and the 3174/3274 cluster controller.

Anyone who has visited an IBM mainframe based customer service center is familiar with clusters of terminal operators connected to the mainframe via one or more cluster controller.

The original 3270 series of synchronous terminal devices was introduced by IBM in 1971. The 3274 cluster controller was introduced in 1977 and worked with the 3278 display terminal. In 1983, IBM released the 3178 display terminal and enhanced the 3274 control unit. The 3174 cluster controller was announced in 1986.

IBM 3270 Emulation

Development lead times in IBM mainframe shops are typically quite long and significant expense is associated with development projects. For these reasons, most CT implementations use an interface known as terminal or cluster controller emulation whereby the CT system appears as just another cluster controller hanging off the mainframe.

The major benefit of this approach it that the CT system makes use of the screens and programs that were developed for terminal operators.

Using the 3270 Presentation Space

An application program (such as CT) and the mainframe (host) communicate through the 3270 presentation space (terminal display). This communication process is known as 3270 terminal emulation.

The Presentation Space (PS)

The 3270 emulation program supports a screen size (screen model) referred to as model 2. See figure 18-1 (not to scale). This space represents 24 rows of 80 columns, for a total of 1920 cells (bytes)

This space does not include the status line (also called the operator information area). The status line is a line at the bottom of each screen that shows the status of the session and information concerning the presentation space. The status line is not considered part of the presentation space and is usually not available to CT application developers.

A separate PS exists for each session (Logical Unit or Device) established with the host. 32 sessions per controller is the typical limit.

Presentation Space for an IBM Model 2 Screen

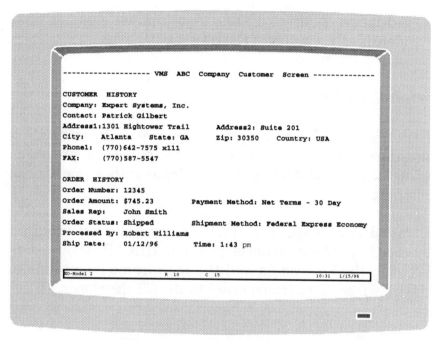

```
------------------- VMS ABC Company Customer Screen --------------

CUSTOMER HISTORY
Company: Expert Systems, Inc.
Contact: Patrick Gilbert
Address1:1301 Hightower Trail      Address2: Suite 201
City:    Atlanta   State: GA       Zip: 30350   Country: USA
Phone1:  (770)642-7575 x111
FAX:     (770)587-5547


ORDER HISTORY
Order Number: 12345
Order Amount: $745.23          Payment Method: Net Terms - 30 Day
Sales Rep:    John Smith
Order Status: Shipped          Shipment Method: Federal Express Economy
Processed By: Robert Williams
Ship Date:    01/12/96         Time: 1:43 pm
```

| EO-Model 2 | | R 10 | C 15 | | 10:31 1/15/96 |

Figure 18-1 (Not to Scale)

Connection Definitions

Before continuing, there are three terms that need to be defined:

- **Device:** A device is the physical piece of equipment that will be attached to the host - a dumb terminal, a host printer or a PC or PC printer using emulation.

- **Session:** A session is a unique set of communications between a device and the host system.

- **Address:** Address refers to a unique number that identifies a session to the host.

What is HLLAPI ?

HLLAPI (High-Level Language Application Programming Interface) allows developers to write application programs using 3270 emulation to communicate with a mainframe computer.

Detailed knowledge of HLLAPI is generally not required by the CT application developer. The HLLAPI layer exists between the CT system and the mainframe host to make the interface more transparent to the CT developer.

Sending Data to the Mainframe

When the CT application has data to send to the mainframe, it fills in the appropriate fields in the screen buffer much like a terminal operator sitting at a keyboard. When the data is complete, the CT application sends the <Enter> key or some other Attention IDentification (*AID*) key. The AID initiates communication with the mainframe which receives and processes the request.

Waiting for the Mainframe to Respond

Once data has been transmitted, the CT application waits for a response from the mainframe. If no response is received in a given time period, the CT application could re-transmit the request, continue to wait or mark the host session as "down" and provide the caller with other choices.

Long response times are typical in heavily loaded mainframe environments. The CT developer may want to play a relatively long message to the caller such as, "Please hold while we process your request." [2 second pause] "Due to heavy activity, we are experiencing some minor delays." [2 second pause, repeat message]. When the host responds, the message in progress will be interrupted and processing can continue. CT application developers call this "dancing".

Receiving Data from the Mainframe

When the mainframe responds, the response is in the form of a screen which will be in the CT application memory. There will be one screen buffer for each session with the mainframe. The CT application can copy data from the appropriate row/column positions into local variables and deliver the information to callers or use it for additional processing.

Connecting to the IBM Mainframe

A CT system usually appears to an IBM host (such as a 3090) as a cluster controller (i.e. 3174 or 3274). The cluster controller is typically connected to a mainframe front end processor (i.e. 3705 or 3725). The other side of the cluster controller attaches to a maximum of 32 IBM terminals. The IBM terminals emulated by the cluster controller in the CT system are 3278 model 2 terminals. Therefore, the CT system attaches to the IBM front end processor and emulates the cluster controller AND each of the terminals behind that controller. The CT system can communicate either SNA/SDLC or Bi-Synchronous to the IBM host. See figure 18-2.

CT System to IBM Mainframe Link

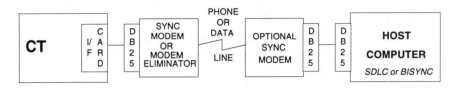

Figure 18-2

The following parameters require information from the MIS department about the particular mainframe with which the application will communicate:

- XID (in hex format)

- Appropriate encoding parameter NRZ or NRZI

- Station (Polling) Address (in decimal format)

- Line mode (Half or Full Duplex)

Log-on Messages and Responses

Most host computers require a certain set of messages be sent to them, to which they make appropriate responses, before they are ready to begin normal data transactions. These messages and their replies are referred to as "log-on" commands and responses and are given as configuration information for each host line.

Log-off Messages and Responses

Similarly, once normal data transactions are completed, a set of messages may be needed to be sent to the host to leave the host in the correct, inactive state; these are referred to as "log-off" messages and, along with their responses, are specified in the host line configuration.

Keep Alive Messages and Response

A third case of special host messages and responses are "keep alive" messages. In some environments, a "keep alive" message is sent to the host during periods of inactivity to keep the host line "alive" i.e. ready for normal transactions and to prevent automatic log-off by the host computer system.

Along with the "keep alive" message and response is a time limit which specifies how often a host transaction should occur to keep the line up and ready.

Time-Outs

To insure that the CT system will not wait indefinitely on a host computer response, a waiting period, given in seconds (the time-out interval), is specified. If the host system does not respond in this amount of time to any message sent to it a special response such as all '?'s is returned to the CT application. When the application determines this response has been returned for a host message, it can take appropriate action such as forwarding the call to a human operator.

Also a number of retries can be specified in the case of host time-outs in which case the last message is sent again for each retry. A CT application wide default action can be specified for host errors/time-outs which will automatically be used if a host transaction fails. This default can be overridden for any particular host transaction that might need special error handling.

Sending 3270 Host Messages

Data transfer to the IBM host is accomplished as follows:

- The screen image must be changed to reflect whatever information needs to be sent to the host.

- An appropriate *AID* (Attention IDentification) generating key must be sent.

To accomplish this the application must "make" the same keystrokes that a terminal operator would use. The keystrokes are encoded by a scheme developed by IBM and widely used for IBM terminal emulation. These key representations are called IBM HLLAPI *mnemonics*.

Keystroke Representation for IBM 3270 Terminals		
HLLAPI Mnemonic	**AID Key**	**Key-Stroke Function**
@A		Alternate (used in conjunction with other keys)
@A@F		Erase Input (@A shifts to Alternate)
@B		Backtab
@C	*	Clear
@D		Delete
@E	*	Enter
@F		Erase EOF (Clear to End of Field)
@I		Insert
@L		Cursor Left
@N		New Line
@R		Reset
@S		Shift (change other keys with this Shift)
@S@y		Field Mark (@S Shifts next key)
@T		Tab
@U		Cursor Up
@V		Cursor Down
@Z		Cursor Right
@0		Home
@1 - @9	*	PF1 - PF9
@a - @o	*	PF10 - PF24
@x - @z	*	PA1 - PA3

Figure 18-3

For character keys like "A", "b", "1", and "#", just the character is specified. For special keys that represent functional actions such as "Tab", "Home", and "PF1", the "@" character is used to indicate that the following character is to represent a special key. The following character would be a "T" for Tab, "C" for Clear, "E" for enter etc. All HLLAPI mnemonics start with "@". Figure 18-3 lists the full set of HLLAPI mnemonics.

Using HLLAPI Mnemonic Keycodes

With HLLAPI, you can emulate a terminal operator who enters the following key sequence:

<Home><Tab><Tab>123<Tab>456<Enter>

with the following string of characters:

@0@T@T123@T456@E

The number "123" is entered into the third unprotected field and "456" into the fourth unprotected field. The AID key "@E" initiates transmission of the data as would the "Enter" key on a terminal. (The code "@0" represents the "Home" key.)

The AID generating key must always be at the end of a series of key strokes since the AID key initiates the data transmission. Any data that follows the AID key is discarded. Therefore, if 2 AID keys need to be sent, the first AID key is sent, followed by a wait for Host Event, then the second AID key is sent. The host will always respond after an AID key is sent.

Tips to Using HLLAPI Mnemonic Keycodes

One common error for first time 3270 CT developers is to perform a clear screen followed by a data entry. For instance, if the developer sends "@C123@E" to clear the screen, type 123, and press <Enter>, only the clear screen is done. This is because the clear screen key is an AID key. This sequence must be done as two host transmissions and two Host Events.

The previous tip requires that two host transmissions and two Host Events be done. Depending on the speed of the IBM host, this could take several seconds for both host transactions to be performed. In many cases, rather than clearing the entire screen before entering a value, the actual field is the only thing that needs to be cleared. The "Clear to End of Field" (@F) mnemonic is NOT an AID key and will not initiate a transmission. Therefore, the previous example could have been coded:

@F123@E

and only one host transmission is required. Depending on the host application, this technique may not work in all cases. However, when it can be utilized, it can save caller time and minimize host activity... both of which are desirable.

Direct Screen Writes

In some special circumstances data from the CT system must be written to specific screen positions to modify the screen image. In this case, the developer may specify the key-stroke to generate the appropriate AID.

243

This technique is much more complex and should only be used when HLLAPI mode is not functioning.

An example of this technique would be if one desired to place the string "123" at row 12 column 13, the string "456" at row 21 column 24 then generate an AID using PF1. A character string to accomplish this would be:

@OR12C13(123@)@OR21C24(456@)@1

The "@OR12C13(123@)" could be read as, "at the offset for row 12 column 13 put the string 123". The next section "@OR21C24(456@)" puts "456" at row 21, column 24. The final "@1" is the code to generate a PF1 AID key.

An option to specifying a location by row/column is to give just the offset into the screen buffer where the offset equals the row minus 1 times 80 plus the column. This would be specified as for the above example by:

@O892(123@)@O1623(456@)@1

The screen positions given for this direct screen modification scheme must be in unprotected areas of the screen. If the string of characters is longer than the unprotected field, then normal skipping/non-skipping to the next unprotected field is performed.

Receiving 3270 Host Messages and Screens

The data received by the CT system is actually a complete copy of the screen image which would be present on a terminal screen. The screen always consists of 24 rows by 80 columns of characters. Attribute bytes that denote fields, blinking characters, etc. may be present in the host data, depending on the application.

3270 host data can be analyzed by the following methods:

- Row/Column Positions

- Unique Text Match

Row/Column Positions

Data can be located by row/column positions or direct offsets into the screen image. Rows are stored consecutively so that column 1 of row 2 is at a direct offset of 82 characters.

Unique Text Match

The data composing the screen image can be searched for specific character strings or something unique which marks the location of desired data. In the order status example, the application could search the screen for the word "STATUS".

Accessing IBM Mainframe Data in the Order Status Application

Cleo DataTalker from Interface Systems, Inc.

The Cleo DataTalker is an IBM 3270 interface board that emulates an IBM 3174/3274 cluster controller handling up to 32 simultaneous 3278 terminal sessions. The DataTalker supports IBM HLLAPI so that communication between the CT system and the mainframe is relatively transparent.

EASE IBM 3174/3274 Emulation

Figure 18-4 shows the tab interface for configuring host communication with EASE.

EASE 3174/3274 Configuration

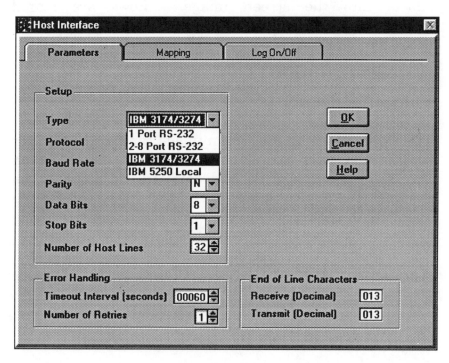

Figure 18-4

Selecting the interface Type, automatically imports default settings for protocol, baud rate, parity, data bits, stop bits and number of host lines. Of course you may override these default setting.

In the bottom left, default error handling is set for a one minute response Time-out and the CT system will automatically re-transmit one time before marking the host session down.

Clicking on the Mapping tab at the top of figure 18-4 will allow you to map phone lines to host session. In an IBM mainframe environment, phone lines and host sessions are usually mapped one-to-one.

Clicking on the Log On/Off tab will allow you to specify:

• Log-on script and expected response(s).

• Log-off script and expected response(s).

• Keep alive messages and expected response(s), if required.

Figure 18-5 shows the development process of building a host request. Items are selected from a pop-up list and dialog boxes walk you through the definition of all data elements required for a particular action.

In the top left corner of figure 18-5, you can see that the caller is prompted to enter their order number as 5 touchtone digits.

Building a Host Request

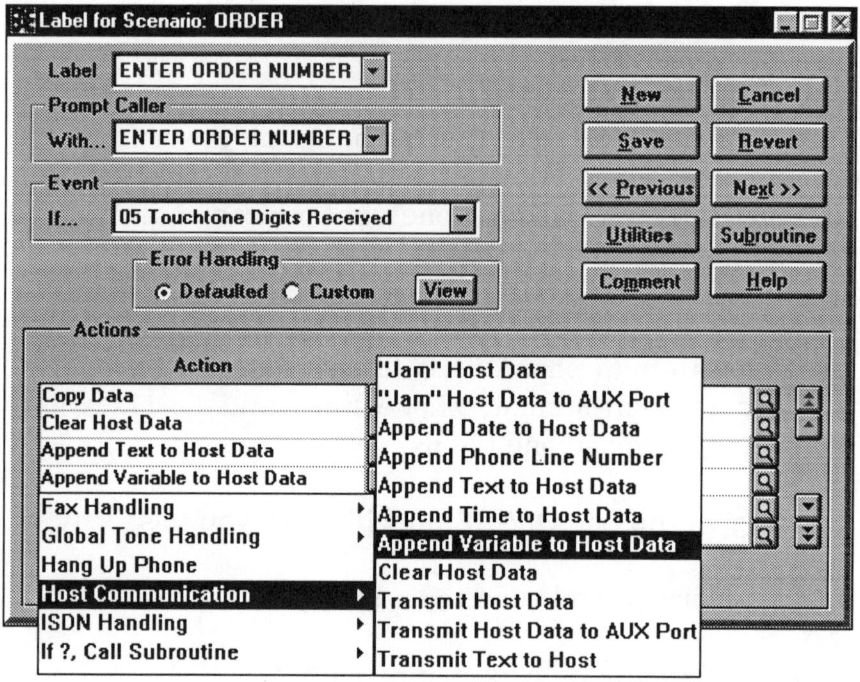

Figure 18-5

Figure 18-6 shows the completed host request. Reading the Actions from top to bottom, you see that the touchtone entry is copied from the Phone Data buffer to a variable called Order Number. The Host Data buffer is cleared. The HLLAPI mnemonics "@0@T@T@T@T@T@T@T @T@T" translate to "Home, Tab, Tab, Tab, Tab, Tab, Tab, Tab, Tab, Tab, Tab, Tab" which will position the cursor on the 11th unprotected field on the 3270 screen (see figure 18-1 for screen layout).

The variable Order Number is inserted in this field. The HLLAPI mnemonic "@E" translates to "<Enter>" which is an IBM Attention IDentification (AID) key. The transaction is now complete and ready to be transmitted to the host for processing.

Transmitting a Host Request

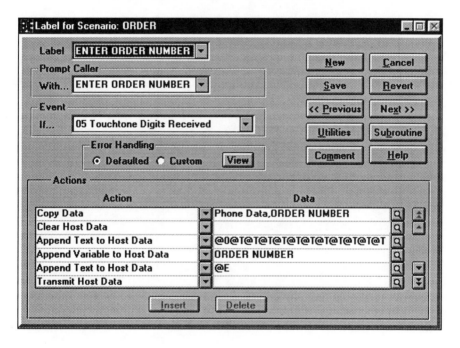

Figure 18-6

19|

Accessing IBM System/3X and AS/400 Data

IBM Midrange Systems

There are four IBM midrange systems:

- AS/400

- System 38

- System 36

- System 34

The System 38 and 34 are fairly rare these days but you may run into them. Smaller companies have found that the System 36 fulfills their needs or they don't have the budget to replace it with the newer AS/400, so there are a number of 36s still in use.

The AS/400 is by far the most prevalent of the IBM midrange systems. There are several models, ranging from the size of a large closet to smaller than a mini-tower PC. Each system is going to be different when you go to a midrange site, depending on the model and additions.

The good news is, each site will have a host administrator who is typically very knowledgeable and probably somewhat possessive about their system. The host administrator will want to deal personally with any host communication issues.

Devices that can be attached to these hosts include dumb terminals and host printers. PCs and PC printers can also communicate using an emulation package. Emulation allows a PC to look and act like one or more dumb terminals to the host system.

IBM 5250 Product Series

IBM designates its mainframe and midrange product lines by the type of dumb terminal that attaches to them. A dumb terminal is basically a monitor and keyboard that connect to the host system and allow access and input to data and programs stored on the host. These devices have no processing power of their own. Disconnected from their host system, they can do nothing.

3270 and 3780 terminals connect to mainframes. The 5250 line is the midrange line of dumb terminals. The most prevalent 5250 terminals are:

- 5251 model 11

- 5292 color

- 5292 graphic

- 5291

Connection Definitions

There are a number of ways that devices can be attached to the host. The most important connections for computer telephony purposes are local attachments to a twinax port.

Before continuing, there are three terms that need to be defined:

- **Device:** A device is the physical piece of equipment that will be attached to the host - a dumb terminal, a host printer or a PC or PC printer using emulation.

- **Session:** A session is a unique set of communications between a device and the host system.

- **Address:** Address refers to a unique number that identifies a session to the host.

Local Connections

Their can be any number of twinax ports on a system and ports can be added at any time. Any twinax port, on any midrange host system, allows for seven addresses. These addresses are numbered 0 through 6, and are available to any device. Local connections to a twinax port can be done with two types of cabling:

- Twinaxial Cable

- Twisted Pair

Twinaxial cable consists of two insulated conductors inside a common insulator, covered by a metallic shield, and enclosed in a cable sheath. Twinax is more stable and more expensive. Devices run with twinax cabling are daisy chained one after another using T-connectors.

Twisted pair uses two insulated copper wires twisted around each other to reduce induction interference from one wire to the other. Twisted pair is the normal cabling from the phone company to an office and from a PBX to a phone. Twisted pair is cheaper, and is used in conjunction with a patch panel. Twisted pair cannot be daisy chained; each terminal requires it's own length of twisted pair.

Log-on Messages and Responses

Most host computers require a certain set of messages be sent to them, to which they make appropriate responses, before they are ready to begin normal data transactions.

These messages and their replies are referred to as "log-on" commands and responses and are given as configuration information for each host line.

Log-off Messages and Responses

Similarly, once normal data transactions are completed, a set of messages may need to be sent to the host to leave the host in the correct, inactive state; these are referred to as "log-off" messages and, along with their responses, are specified in the host line configuration.

Keep Alive Messages and Response

A third case of special host messages and responses are "keep alive" messages. In some environments, a "keep alive" message is sent to the host during periods of inactivity to keep the host line "alive" i.e. ready for normal transactions and to prevent automatic log-off by the host computer system. Along with the "keep alive" message and response is a time limit which specifies how often a host transaction should occur to keep the line up and ready.

Time-Outs

To insure that the CT system will not wait indefinitely on a host computer response, a waiting period, given in seconds (the time-out interval), is specified. If the host system does not respond in this amount of time to any message sent to it a special response such as all '?'s is returned to the CT application. When the application finds this response has been returned for a host message, then it can take appropriate action such as forwarding the call to a human operator.

Also a number of retries can be specified in the case of host time-outs in which case the last message is sent again for each retry. An application wide default action can be specified for host errors/time-outs which will automatically be used if a host transaction fails. This default can be overridden for any particular host transaction that might need special error handling.

Sending 5250 Host Messages

Data transfer to the IBM midrange host is accomplished as follows:

- The screen image must be changed to reflect whatever information needs to be sent to the host.

- An appropriate *AID* (Attention IDentification) generating key must be sent.

To perform these two steps the CT system should specify the same keystrokes that a terminal operator would use. The keystrokes are encoded by the following scheme developed by IBM and widely used for IBM terminal emulation.

These key representations are called IBM HLLAPI mnemonics. HLLAPI stands for High Level Language Application Program Interface. HLLAPI mnemonics were actually not defined for 5250 terminals, but this representation is used for consistency across 3270 and 5250 interfaces.

For character keys like "A", "b", "1", and "#", just the character is specified. For special keys that represent functional actions such as "Tab", "Home", and "PF1", the "@" character is used to indicate that the following character is to represent a special key. The following character would be a "T" for Tab, "C" for Clear, "E" for enter etc. All HLLAPI mnemonics start with "@". The following figure lists the full set of HLLAPI mnemonics as used with 5250 emulation.

Keystroke Representation for IBM 5250 Terminals		
HLLAPI Mnemonic	**AID Key**	**Key-Stroke Function**
@A@F		Erase Input (@A shifts to Alternate)
@B	-	Backtab
@C	*	Clear
@D		Delete
@E	*	Enter
@F		Erase Input
@G	*	Roll Down
@H	*	Help (AID if non error mode)
@I		Insert
@L		Cursor Left
@N		New Line
@Onnnn (xxx@)		Places the xxx data at Offset nnnn
@ORnnCmm (xxx@)		Places the xxx data at Row nn and Column nn
@P	*	Roll Up
@Q		Attn (attention)
@R		Error / Reset
@T		Tab
@U		Cursor Up
@V		Cursor Down
@W		Backspace
@X		Field Exit
@Y		System Request
@Z		Cursor Right
@0	*	Home (AID generated if cursor already at Home position; AID causes last screen to be resent.)
@1 - @9	*	F1 - F9
@a - @o	*	F10 - F24
@X		Field Exit

Using HLLAPI Mnemonic Keycodes

With HLLAPI, you can emulate a terminal operator who enters the following key sequence:

<Home><Tab><Tab>123<Tab>456<Enter>

with the following string of characters:

@0@T@T123@T456@E

The number "123" is entered into the third unprotected field and "456" into the fourth unprotected field. The AID key "@E" initiates transmission of the data as would the "Enter" key on a terminal. (The code "@0" represents the "Home" key.)

The AID generating key must always be at the end of a series of key strokes since the AID key initiates the data transmission. Any data that follows the AID key is discarded.

Receiving 5250 Host Messages and Screens

The data received by the CT system is actually a complete copy of the screen image which would be present on a terminal screen. The screen always consists of 24 rows by 80 columns of characters.

Attribute bytes that denote fields, blinking characters, etc. may be present in the host data, depending on the application.

5250 host data can be analyzed by the following two methods:

- Row/Column Positions

- Unique Text Match

Row/Column Positions

Data can be located by row/column positions or direct offsets into the screen image. Rows are stored consecutively so that column 1 of row 2 is at a direct offset of 82 characters.

Unique Text Match

The data composing the screen image can be searched for specific character strings or something unique which marks the location of desired data. In the order status application, you could search for the word "STATUS".

Receiving 5250 Host Errors

If the host responds with an error message, the keyboard is usually locked until the error is acknowledged. No other transmissions can be made until the error is cleared. The error/reset key can be sent to clear the error and unlock the keyboard.

Accessing IBM System 3X and AS/400 Data in the Order Status Application

Micro-Integration's 5250 Local Card

Micro-Integration's 5250 Local card is an IBM 5250 interface board that emulates up to seven IBM 5251 Model 11 terminal sessions simultaneously.

The 5250 interface board is typically connected to the mid-range host via twinax cable.

EASE 5250 Emulation

Figure 19-2 shows the tab interface for configuring a host interface with EASE.

The Mapping tab allows you to map phone lines to host sessions. In this example the phone lines are mapped one-to-one to host sessions. However, the 5250 emulation is limited to 7 sessions.

In larger systems, such as a T-1 based 24 line system, you may want to take advantage of the Host Pooling feature. This would allow you to map the first four phone lines to host session 1, the second four lines to session 2, and so on.

When using Host Pooling, you may need to get the MIS department to consolidate the information from multiple 5250 screens in to a single "super screen". This will greatly improve the CT system through-put.

EASE 5250 Configuration

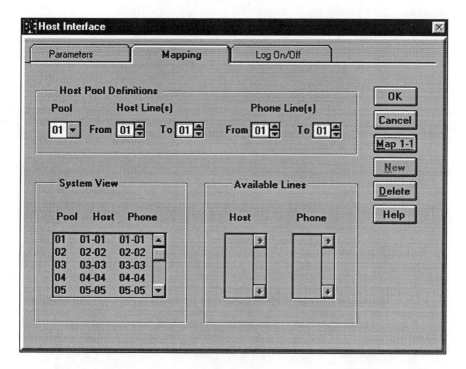

Figure 19-2

Figure 19-3 shows how data is retrieved from a host response screen and stored in variables for the CT system to use. The dialog box in the top right indicates that the cursor will be positioned at row 10 column 20, 10 characters will be retrieved and stored in a variable called Order Status.

As the Actions at the bottom of the screen show, this process is repeated at different row/column coordinates until the order amount and order ship date are also retrieved. After all three variables are collected, the CT system proceeds to process the caller's request.

Processing a Host Response

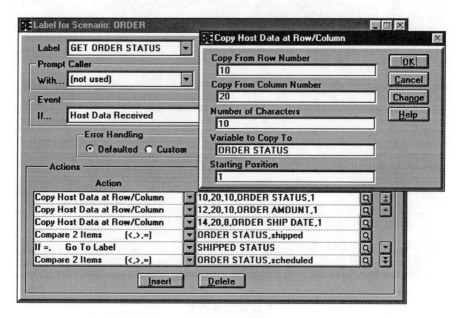

Figure 19-3

20|

Accessing Data Via RS-232 Serial Ports

Fundamentals of RS-232 Communication

Evolution of the RS-232 Standard

RS-232 is currently the most common data communication interface and signaling standard. The Electronic Industries Association published EIA Standard RS-232-C in August of 1969 under the name "Interface Between Data Terminal Equipment And Data Communication Equipment Employing Serial Binary Data Interchange". RS-232-C is based on RS-232-B, was modified to RS-232-D in 1987 and the "E" variant is currently in the works. RS-232-C is an important point in the evolution primarily because of market timing and secondarily because of what was not dictated in the standard.

The 1970s was the boom period for remote computing/time-sharing computer service bureaus. RS-232-C flourished during this period because it matched the current needs of this time - dumb terminals talking to remote computers over modems. The standard defined the interface between *Data Terminal Equipment* (DTE) and *Data Communication Equipment* (DCE) over telephone networks. However, RS-232-C did not define a physical interface connector and it did not contemplate the PC

263

revolution. These short-comings encouraged the creativity of device manufacturers who saw a general purpose specification that could be bent to their particular needs.

One of the biggest leaps in the evolution of RS-232-C came with its adaptation as a general purpose serial input/output port for PCs and locally attached devices. Over time, the definition of DTE has expanded to include computers, especially PCs, and not just CRTs or other terminals.

Likewise, the DCE can be a modem, but it can also be any device that responds like a DCE. This evolution is the primary reason that the RS-232-D and RS-232-E initiatives were launched, and also explains why RS-232 communication often involves a soldering iron and numerous cable change-outs.

Physical Interface

As mentioned earlier, one of the big surprises of the RS-232-C standard is that it does not address the interface connector. In 1969, on the last page of the specification, the drafters state, "While no industry standard exists which defines a suitable interface connector, it should be noted that commercial products are available which will perform satisfactorily as electrical connectors for interfaces specified in RS-232-C..." With this closing of the standard, the drafters left this issue open to the imagination of device manufacturers until 1987.

With the release of RS-232-D, the Electronic Industries Association formally recognized the *DB-25* connector which by that time had become almost universally identified with RS-232-C. Even then the issue was not resolved because IBM and some other manufacturers use a smaller *DB-9* connector.

Male and Female Connectors

The RS-232-C standard specifies that the male connector is associated with the DTE and the female connector is associated with the DCE.

DB-25 Connectors

DB-25 connectors faithfully lay out the pins for each of the 25 circuits defined in the original RS-232-C standard. Figure 20-1 shows the pin layout and assignments.

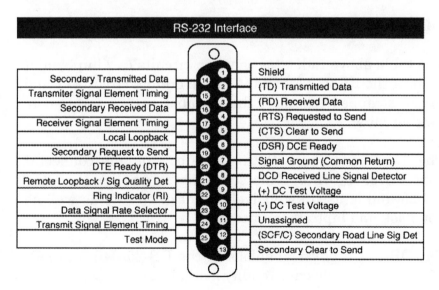

Figure 20-1

Favorite RS-232 Pin Pairs

Most important data communication functions in the RS-232 standard can be executed and controlled over five pairs of circuits, such as transmit/receive. As a reference convention, the RS-232 standard views all protocol issues from the DTE standpoint so that transmit means that the DTE transmits. The following table identifies the most important pin pairs in the logical order of a data transmission:

Pin Pair	Circuit Functions
8/22	Carrier Detect/Ring Indicator
6/20	Data Set Ready/Data Terminal Ready
4/5	Request to Send/Clear to Send
2/3	Transmitted Data/Received Data
1/7	Protective Ground/Signal Ground

DB-9 Connectors

The table above indicates that ten circuits can implement most of the RS-232 standard. If the ground circuits (1 and 7) are strapped together, this number can be reduced to nine circuits.

The DB-9 does just that. It is a nine-pin connector popularized by IBM. This connector is smaller than the DB-25 and is used extensively on desktop PCs as well as notebook and laptop computers.

The DB-9 is a physical subset of the DB-25. DB-9 to DB-25 converter cables are readily available.

The following table illustrate the DB-9 to DB-25 pin mapping:

DB-9	Circuit Function	DB-25
1	Carrier Detect	8
2	Received Data	3
3	Transmitted Data	2
4	Data Terminal Ready	20
5	Signal Ground	7
6	Data Set Ready	6
7	Request to Send	4
8	Clear to Send	5
9	Ring Indicator	22

Three Pin Connections

RS-232 communication may be accomplished using just three of the pins: pin 2 (transmit), pin 3 (receive) and pin 7 (ground). This is considered "open loop" or wide-open RS-232 communication with no safety net. This is usually accomplished by simply clipping or de-soldering wires from standard connectors.

Ones and Zeros: Transmitting and Receiving Data Bits

Under RS-232-C, a positive voltage in the range of 5 to 15 volts on pin 2 (transmit) or pin 3 (receive) with respect to pin 7 (ground) represents a "0"; and, a negative voltage between -5 and -15 volts represents a "1". The RS-232-D standard extends this range to +25 volts and -25 volts respectively.

Null Modem for DTE to DTE Configurations

As dumb terminals evolved into intelligent PCs, it became advantageous to move beyond the DTE to DCE configuration originally contemplated in the RS-232 standard. Inventive people realized that DTE to DTE and DCE to DCE communication was possible by simply crossing pins 2 and 3 (receive and transmit). When using RS-232 communication in an locally connected computer telephony implementation, DTE to DTE is the most popular approach.

DTE to DTE configuration conversion can be made with a readily available *null modem cable* or *modem eliminator*; however, it is advisable to obtain a wiring diagram. Although this practice is widely used, it is a non-standard implementation. Many man-days have been spent trouble-shooting cables and many problems have been solved by re-soldering connectors.

Limitations of RS-232 Communications

Currently RS-232 is the most common data communications standard. It has reached this status over years of evolution and adaptation. However there are some limitations to consider:

- The distance between DTE and DCE is usually limited to a maximum of 50 feet, due to transmission line loss. This is a conservative figure and is greatly dependent on the type of cable, grounding, and transmission power.

- The transmission speed is usually limited to 19,200 bits per second.

- Grounding problems can cause data errors because the ground is used as a reference point for interpreting data bits.

- Data flow control must be handled independently from the RS-232 data communication.

Connecting the RS-232 Host Computer Link

For single port RS-232 asynchronous communications to a host computer or peripheral device, the CT system uses a standard PC serial port such as COM2. Some PC manufacturers refer to serial ports as serial I/O connectors.

Some PC manufacturers use a DB9 (9 pin connector) instead of a DB25 (25 pin connector) for COM2. If the COM2 connector is a DB9, a DB9 to DB25 conversion cable may be required.

Null Modem versus Modem RS-232 Links

The RS-232 interface may be one of two types:

- Null Modem Cable

- Modem

A null modem cable typically is used to connect to a host computer which is located within 50 feet of the VRU. A null modem cable is a standard 25 or 9 pin RS-232 cable that has the number two (transmit) and number three (receive) pins swapped.

Null modem inserts are also available which enable two standard RS-232 cables to be connected as a single null modem link.

An asynchronous modem is typically used to connect to a host computer which is more than 50 feet from the VRU or which must be accessed via a public or leased phone line. This modem can be 300, 1,200, 4,800, 9,600 or 19,200 bps, but the asynchronous modem on the host computer's end must be the same baud rate and type as the VRU.

Null Modem RS-232 Link

Figure 20-2 shows an RS-232 link using a null modem.

Null Modem RS-232 Link

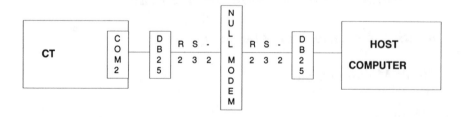

Figure 20-2

Modem RS-232 Link

Figure 20-3 shows an RS-232 link using modem connections.

Modem RS-232 Link

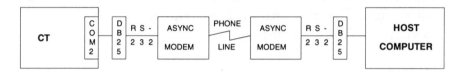

Figure 20-3

Log-on Messages and Responses

Most host computers require a certain set of messages be sent to them, to which they make appropriate responses, before they are ready to begin normal data transactions. These messages and their replies are referred to as "log-on" commands and responses and are given as configuration information for each host line.

Log-off Messages and Responses

Similarly, once normal data transactions are completed, a set of messages may be needed to be sent to the host to leave the host in the correct, inactive state; these are referred to as "log-off" messages and, along with their responses, are specified in the host line configuration.

Keep Alive Messages and Response

A third case of special host messages and responses are "keep alive" messages. A "keep alive" message is sent to the host during periods of inactivity to keep the host line "alive" i.e. ready for normal transactions and to prevent automatic log-off by the host computer system. Along with the "keep alive" message and response is a time limit which specifies how often a host transaction should occur to keep the line up and ready.

Time-Outs

To insure that the CT system will not wait indefinitely on a host computer response, a waiting period, given in seconds (the time-out interval), is specified. If the host system does not respond in this amount of time to any message sent to it a special response such as all '?'s is returned to the CT application. When the application finds this response has been returned for a host message, then it can take appropriate action such as forwarding the call to a human operator.

Also a number of retries can be specified in the case of host time-outs in which case the last message is sent again for each retry. A scenario wide default action can be specified for host errors/time-outs which will automatically be used if a host transaction fails. This default can be overridden for any particular host transaction that might need special error handling.

RS-232 in the Order Status Application

Protocol Overview

This section describes a typical transactional protocol that would be used between a CT system and a Tandem NonStop computer system to implement the order status CT application.

The basic protocol structure could apply to communication with another PC or even a PBX that supports RS-232 command and control.

Communication Link

The physical link between the CT system and the Tandem NonStop computer shall be:

- RS-232-C

- Asynchronous

- Local, Null Modem, DTE to DTE

- 19,200 bps

- Even Parity

- 8 Data Bits

- 1 Stop Bit

Data Link Communication Protocol

The CT system will use transactional ASCII protocol. Every transaction from the CT system will require a response from the Tandem before the CT system can initiate another transaction. The CT system will disregard any unsolicited response from the Tandem.

Standard format will be a two digit phone line identifier (used like a terminal identifier) followed by a two digit transaction code. Data following the transaction code will be fixed field and dependent on the specific transaction code. Variable length numeric data items will be right justified and left filled with zeros. Variable length alpha data items will be left justified and right filled with spaces. All transactions will be terminated with a carriage return.

The two digit line identifier will specify the phone line currently being processed. Specific line identifiers range from 01 to 48 for each CT system in use. The CT system will use a separate RS-232 link for each 16 phone lines to be handled. Host to host communication which is not phone line specific will use the line identifier 00. The Tandem must return the line identifier with every response.

Transactions from the CT System to the Tandem

00PA
Please acknowledge. Used for sign on and keep alive. No additional data.

01CO to 48CO

Customer and order numbers for status inquiry. The CO code will have 14 numeric characters in the data field. Nine for customer number and five for order number.

01TC to 48 TC

Transferring caller on current phone line XX to customer service. The TC code will have 15 characters in the data field. Nine for customer number, five for order number, and one alpha character for transfer reason code. If no customer number or order number was received from the caller by the CT system, the item will be filled with zeros. Transfer reason codes include:

T - Two no entry time-outs, or two invalid entries.

C - Credit or shipping hold on order.

N - Order number not found.

S - Status reported, transferred at caller's request.

Transactions from the Tandem to the CT System

00OK

CT system is acknowledged as on-line. No additional data.

01SH to 48SH

Order shipped. 14 characters in data field. Six digit ship date and eight digit invoice amount.

01SC to 48SC

Order scheduled for shipment. 14 characters in data field. Six digit shipping date and eight digit invoice amount.

01CA to 48CA

Order canceled. Eight digit invoice amount.

01HL to 48HL
Order on credit or shipping hold. Eight digit invoice amount.

01NF to 48NF
Order not found. No additional data.

01TX to 48TX
Call transfer to customer service confirmed.

EASE Log On Configuration

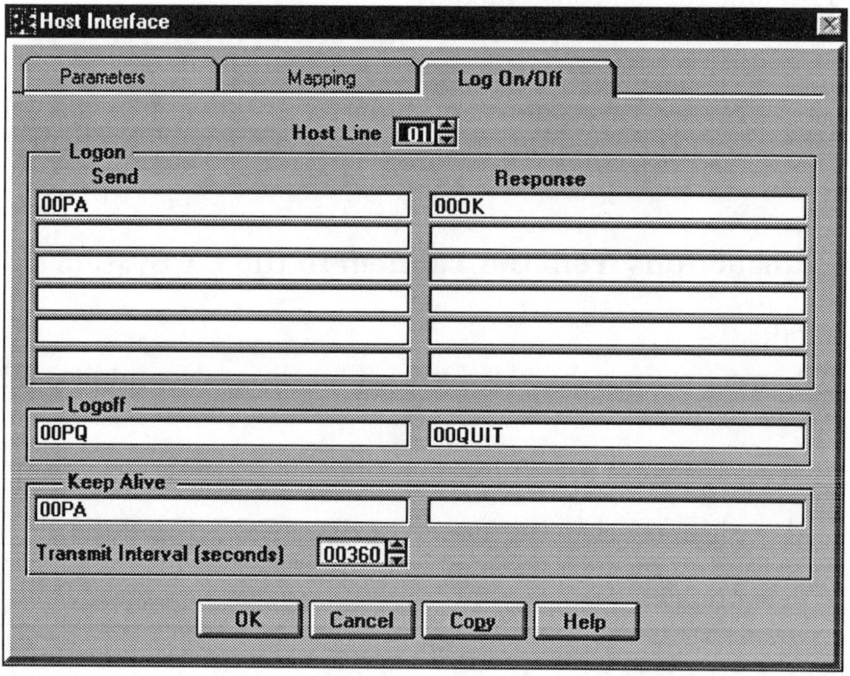

Host Interface	☒

Parameters	Mapping	Log On/Off	

Host Line [01] ⇕

Logon

Send	Response
00PA	000K

Logoff

00PQ	00QUIT

Keep Alive

00PA	

Transmit Interval (seconds) [00360] ⇕

OK	Cancel	Copy	Help

Figure 20-4

EASE RS-232 Interface

Figure 20-4 shows the tab interface for configuring host communication with EASE.

The log-on sequence consists of the CT system sending "00PA" for non-line specific please acknowledge. The host responds with "00OK" for non-line specific okay.

The log-off sequence consists of the CT system transmitting "00PQ" for non-line specific please quit. The host responds with "00QUIT" for non-line specific quit.

The keep alive sequence consists of the CT system transmitting "00PA" for non-line specific please acknowledge every six minutes to keep the host from automatically logging the CT system off during periods of inactivity. The host responds with "00OK" for non-line specific okay.

Retrieving RS-232 Data

Figure 20-5 shows the process of retrieving specific data from the host data buffer and storing it in variables for use by the CT system.

The dialog box in the top right of figure 20-5 shows the CT system looking for an exact match between "SH" for "shipped" in the host data buffer. If it finds an exact match, it stores the position of "SH" as a direct offset (number of characters) into the buffer.

Based on the position of "SH" in the host data buffer, the CT application retrieves the next 14 characters. It then takes the first six characters and copies them into a variable called Order Ship Date. Then it takes the next eight characters and copies them into a variable called Order Amount. The data is now ready for use by the CT application.

Retrieving RS-232 Data

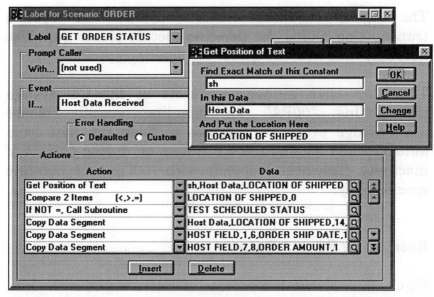

Figure 20-5

21
Accessing Clipper, dBASE and FoxPro Data

Fundamentals of Xbase

File Server Architecture

More powerful PCs and the introduction of the local area network (LAN) marked the advent of wide spread distributed processing. PC users can link their PCs together via a hub or file server to store and share programs and information. The file-server computer stores information and sends data files to PCs or workstations as requested.

Database Concepts

A database is simply a collection of information organized in a logical manner. The telephone book is an excellent example of a database. Names, addresses and phone numbers are logically organized.

Telephone Book Example of a Database

Name	Address	Phone #
Adams Painting	3143 Canton Rd	424-1999
Baker Construction	1422 Peachtree St	892-1730
Cagel & Company	3541 Habersham	939-1102
Darcy Sports	3393 Northside Dr	255-2189

In the example above, the telephone book is a data file. Each listing in the phone book is a record and each piece of information in that record is a field.

Database

A logically organized collection of information residing in one or more data files.

Data File

A computer file which stores data. Clipper, dBASE and FoxPro (collectively known as Xbase) traditionally use a ".DBF" file extension on data file names.

Record

Data file information is organized into records. A record is composed of fields (e.g. one name, one address and one phone number). Each record has a unique record number and a deletion flag associated with it.

Record Number

A unique number designating the physical location of a record in a data file.

Deletion Flag

Indicates that a record has been marked for deletion. Records are not physically deleted until the data file is packed.

Packing

Packing physically removes records from the data file based on the deletion flag. Packing can be a time consuming process with large data files. This time constraint is the reason that records are marked for deletion during normal processing but not physically deleted until time allows.

Field

Records are organized into fields. A field is one category of information in each record. The fields in the telephone book example are name, address and phone number. The Xbase .DBF standard uses four attributes to describe each field:

- Name: Identifies the field. The name must be unique within the data file and may only contain alphanumeric or underscore characters up to a maximum of ten characters.

- Type: Identifies the kind of information that will be stored in the field. Field types include character, numeric, date, floating point, logical, memo and general.

- Length: Identifies the maximum number of characters or digits allowed in a field. The length of a date field is fixed at 8, logical at 1 and memo at 10.

- Decimal: Identifies the number of digits after the decimal point in numeric and floating point fields. Other fields are fixed at 0.

Index

The index determines how information in the data file will be logically sorted for presentation. The index is a file which contains index keys for one or more tags.

The Xbase standard provides for four major index file formats which are usually identified by their file name extensions:

- .NTX: Clipper (Computer Associates)

- .NDX: dBASE III and dBASE III Plus (Borland)

- .MDX: dBASE IV (Borland)

- .CDX: FoxPro (Microsoft)

The .NTX and .NDX formats impose a limit of one tag per index file while the .MDX and .CDX formats allow multiple tags in each index file.

Tag

A tag describes the order in which data file records will be accessed. A tag is a logical sorting and does not affect the physical location of records in the data file.

Index Key

The index key is the field or field expression on which the tag's logical sorting is based. In the telephone book example above, the tag is based on name as the index key.

Re-Indexing

An index file may lose synchronization with its corresponding data file if a computer loses power during an update or if the index file is closed or inactive when the data file is modified. When this occurs, it is necessary to re-index the index file to remedy the problem.

Filter

Filters are calculations performed on field information which return true/false conditions. Filters allow fast access to a subset of records in a data file.

Relation

A relation is a logical connection between two or more data files. A relation provides a link from a record in the current data file to related records in other data files.

Performance Considerations

An application may require that several data files be open at the same time. However, there is a performance tradeoff. Opening and closing data files require CPU and disk access time. Regularly used data files should be left open subject to memory constraints because each open data file requires an area of memory called a record buffer.

Multi-User LAN Considerations

In a production environment, multiple people or applications may try to access and modify the same data at the same time. This could result in data corruption if not prevented. The safety measure is known as locking.

Locking is a first-come-first-served method of sharing a database. When an application locks data, it notifies other applications that particular data cannot be modified at the current time.

There are three principal types of locking: data file locking, record locking, and index locking.

Data File Locking and Unlocking

A data file lock prevents all existing records in the data file from being modified by other applications. A file lock also prevents new records from being appended to the current data file.

Record Locking and Unlocking

Record locking prevents an individual record from being modified. Record locking is the lowest level of multi-user safeguards. Individual fields cannot be locked.

Index Locking and Unlocking

Appending records to a data file and certain modifications of records result in the need to lock and update index files. This locking and unlocking typically occurs automatically.

Fatal Embrace

Fatal embrace (also known as fatal lock or deadlock) occurs when two applications try to lock the same items, but do so in reverse order.

The first application initiates a lock and waits on the second application at the same time that the second application initiates a lock and waits for the first application. The result is that both applications are waiting on each other and therefore they both wait forever in a fatal embrace.

Fatal embrace may be easily avoided by simply assuring that all applications always lock data files in the same order. This assures that the first application that locks the first data file will be allowed to complete all necessary locking, processing, and unlocking before the second application receives a successful lock.

Seeking

The principal function of an index is to find a record based on a search key (such as an order number that has been entered). This process is call seeking and consists of comparing the search key with the index keys in a designated tag.

When the seek results in a match, the corresponding record is read into the record buffer. From this point data fields may be accessed and manipulated.

Record Buffer and Current Record

The record buffer is an area in memory that is used to store one record from a data file. This record is referred to as the current record.

Any changes made to the current record must be written to the data file before another current record is read into the record buffer.

Field Functions

From a CT application perspective, field functions retrieve and manipulate fields of the current record in the record buffer. Typical field functions would include Get Field and Put Field.

Accessing Xbase Data in the Order Status Application

Relationship of Xbase Actions

It is important to understand the relationship between all the Xbase actions in order to develop a bullet-proof Xbase CT application.

Xbase Relationships

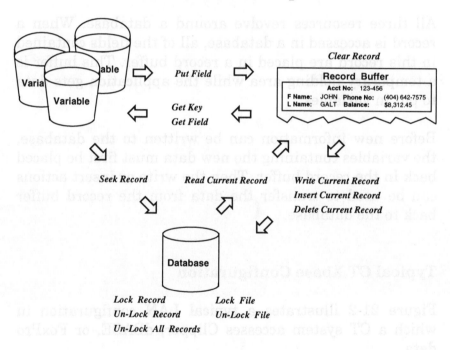

Figure 21-1

All the actions revolve around three resources:

- Databases

- Record Buffers

- Variables

Figure 21-1 illustrates the relationships of all the Xbase actions. Each Xbase action appears in italics. The arrows show the flow of data between resources.

All three resources revolve around a database. When a record is accessed in a database, all of the fields contained in this record are placed in a record buffer. This buffer is a temporary holding area while the application gets data to put in variables.

Before new information can be written to the database, the variables containing the new data must first be placed back in the record buffer. Then the write or insert actions can be used to transfer the data from the record buffer back to the database.

Typical CT Xbase Configuration

Figure 21-2 illustrates a typical LAN configuration in which a CT system accesses Clipper, dBASE, or FoxPro data.

Of course, the LAN is not required. The Clipper, dBASE, or FoxPro databases could reside on the CT systems' local hard disk drive.

CT Configuration with Xbase

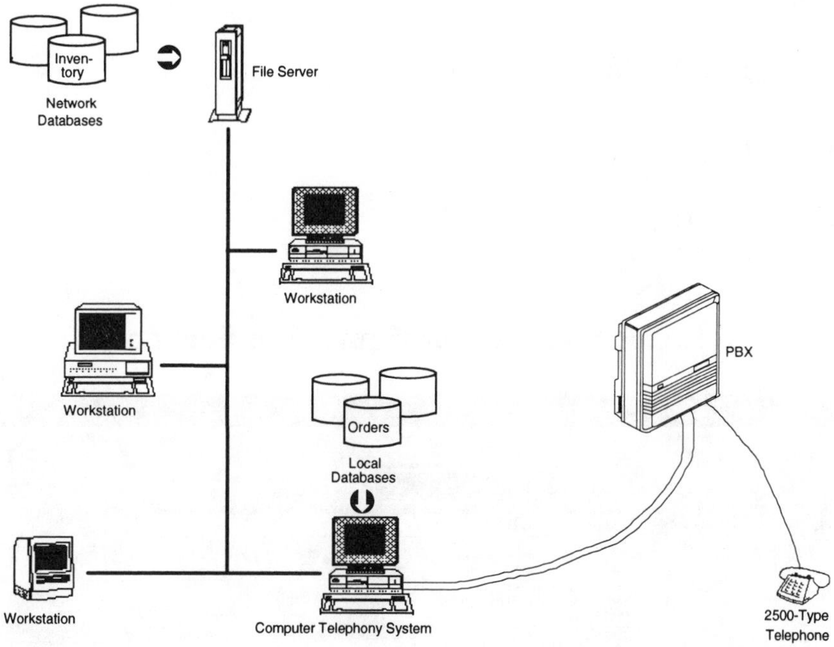

Figure 21-2

EASE Xbase Configuration

Figure 21-3 shows the EASE FoxPro configuration screen. Of course Clipper and dBASE are also supported.

The top of the screen shows that the CT application will use the ORDER.DBF data file and ORDER.CDX index both located on the network F: drive.

Notice in the bottom right of figure 21-3 that available fields are automatically imported into EASE. For fast access, the CT application will use the following index tags:

- O_NUMBER

- O_STATUS

- SHIP_DATE

EASE FoxPro Configuration Screen

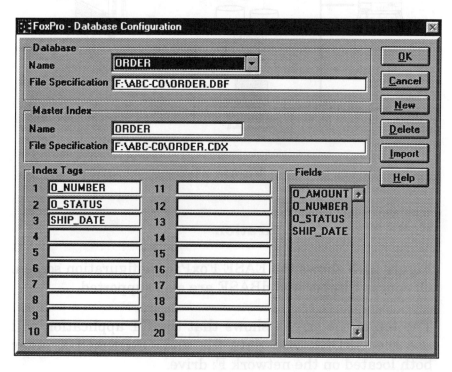

Figure 21-3

EASE Xbase Data Access

Figure 21-4 shows the EASE actions for accessing Xbase data. Notice that record and file locking are supported for data integrity in a multi-user environment.

Xbase Data Access

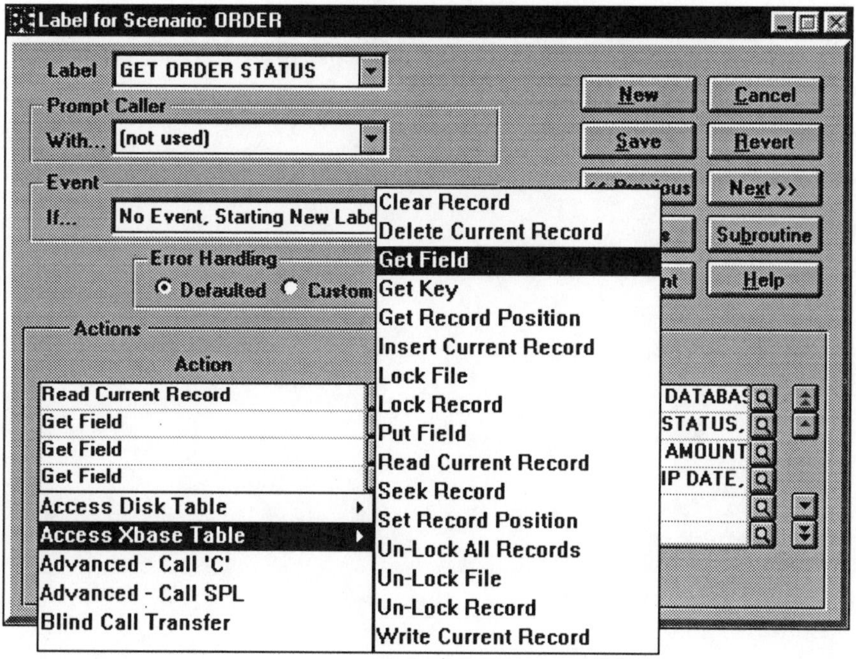

Figure 21-4

In the order status example illustrated in figure 21-4, the data access process involves seeking a record based on the order number tag. The current record is read into the record buffer and then three sequential get field functions are performed. These actions put the appropriate data fields into the variables Order Status, Order Amount and Order Ship Date. The data is now available for use by the CT application.

22

Accessing SQL Data with ODBC

SQL

Client-Server Architecture

Client-server architecture is considered the next evolutionary step beyond file server architecture (discussed in the previous chapter). As the name suggests, client-server configurations are based on a division of labor. On the front end, client workstations run applications; and, on the back end, database servers store data and manage data requests.

The key difference between client-server and the traditional file-server architecture is that the database server processes requests for data (not files) and transmits only the requested data to applications on the client PC. This division of labor allows the client PC to focus on the user interface, user applications and data requests, while the database server focuses on database management for concurrent access, data integrity, security, and error recovery.

The key advantages of client-server architecture can include:

- Efficient division of labor

- Scalability

- Reduced network traffic

- Off-loading of client processors

- Better access to data

- Better data safeguards

- Lower costs than legacy systems

Relational Databases

In the client-server model, most currently available database servers use a relational database in which data is stored in its most natural, fundamental structure; and mathematical relationships provide the basis of access. A relational database consists of tables, indexes, keys, rows, and columns. Tables relate through at least one common column.

The relational database model provides a number of important characteristic that make database use and management relatively easy, error-resistant, and predictable.

The most critical characteristic is the mathematical foundation of the relational model. This precise mathematical basis provides database integrity in somewhat the same way that double-entry accounting forces accounting records to always be in balance.

History of the SQL Language

According to authorities, SQL is pronounced "S Q L" not "sequel"; and due to evolution, the letters no longer stand for anything, even though Microsoft persists in identifying SQL as an abbreviation for "Structured Query Language".

In the mid 1970s, IBM Research Laboratory developed a language to implement the relational database model. The prototype language was called SEQUEL for Structured English Query Language and was the application programming interface to their embryonic relational database called System R. In 1977, a new design prototype emerged which was called SEQUEL/2; but, the name was later changed to SQL. This is the source of the popular belief that SQL is an acronym for Structured Query Language and is pronounced "sequel".

By 1979, the word was out that IBM was going to launch a product based on the SQL language. In classic David and Goliath fashion, a small company named Relational Software (currently known as Oracle) beat IBM to market.... and the rest is history.

Currently, SQL is the accepted standard for data definition, data manipulation, data management, access protection, and transaction control of relational databases. It uses tables, indexes, keys, rows, and columns to identify storage locations.

Many types of applications use SQL statements to access data. Examples include ad hoc query facilities, decision support applications, report generation utilities, and online transaction processing systems.

SQL is not a complete programming language in itself. For example, there are no provisions for flow control. SQL is normally used in conjunction with a traditional programming language.

SQL Statements

SQL causes things to happen by the execution of SQL statements. SQL statements may be divided into three major categories:

- **Data manipulation statements** which retrieve, store, delete, or update data in the database.

- **Data definition statements** which allow you to define and modify the definition of your database.

- **Management statements** which allow you to control the database aspects of an application such as the termination of transactions or the setting of certain parameters that affect other statements.

SQL-89

SQL was first standardized by the *American National Standards Institute* (ANSI) in 1986. The first ANSI standard defined a language that was independent of any programming language.

The most widely implemented standard is ANSI 1989, which defines three programmatic interfaces to SQL:

- Module language: Allows the definition of procedures within compiled programs (modules). These procedures are then called from traditional programming languages. The module language uses parameters to return values to the calling program.

- Embedded SQL: Allows SQL statements to be embedded within a program. The specification defines embedded statements for COBOL, FORTRAN, Pascal, and PL/1.

- Direct invocation: Access is implementation-defined.

The most popular programmatic interface has been embedded SQL.

Embedded SQL

Embedded SQL allows programmers to place SQL statements into programs written in a standard programming language (for example, COBOL or Pascal), which is termed the host language. SQL statements are delimited with specific starting and ending statements defined by the host language. The resulting program contains source code from two languages — SQL and the host language.

When compiling a program with embedded SQL statements, a pre-compiler translates the SQL statements into equivalent host language source code. After pre-compiling, the host language compiler compiles the resulting source code.

SQL-92

SQL-92 is the most recent ANSI specification, and is now an international standard. SQL-92 defines three levels of functionality: entry, intermediate, and full. SQL-92 contains many new features, including support for dynamic SQL.

Dynamic SQL allows an application to generate and execute SQL statements at run-time. When an SQL statement is prepared, the database environment generates an access plan and a description of the result set. The statement can be executed multiple times with the previously generated access plan, which minimizes processing overhead.

Dynamic SQL is not as efficient as static SQL, but is very useful if an application requires:

- Flexibility to construct SQL statements at run-time.

- Flexibility to defer an association with a database until run-time.

Open Database Connectivity (ODBC)

ODBC Overview

Open Database Connectivity (ODBC) is Microsoft's strategic interface for accessing data in a heterogeneous environment of relational and non-relational database management systems. Based on the Call Level Interface specification of the SQL Access Group, ODBC provides an open, vendor-neutral way of accessing data stored in a variety of proprietary personal computer, minicomputer, and mainframe databases.

ODBC alleviates the need for independent software vendors and corporate developers to learn multiple application programming interfaces. ODBC now provides a near universal data access interface. With ODBC, applications can concurrently access, view, and modify data from multiple, diverse databases. ODBC is a core component of Microsoft *Windows Open Services Architecture* (WOSA).

Providing data access to applications in today's heterogeneous database environment is very complex for software vendors as well as corporate developers. With ODBC, Microsoft has eased the burden of data access by creating a vendor-neutral, open, and powerful means of accessing *database management systems* (DBMSs).

- ODBC is vendor neutral, allowing access to DBMSs from multiple vendors.

- ODBC is open. Working with ANSI standards, the SQL Access Group (SAG), X/Open, and numerous independent software vendors, Microsoft has gained a very broad consensus on ODBC's implementation, and it has become the dominant standard.

- ODBC is powerful—it offers capabilities critical to client/server on-line transaction processing (OLTP) and decision support system (DSS) applications, including system table transparency, full transaction support, scrollable cursors, asynchronous calling, array fetch and update, a flexible connection model, and stored procedures for "static" SQL performance.

ODBC Benefits

ODBC provides many significant benefits to developers, end users, and the industry by providing an open, standard way to access data.

- ODBC allows users to access data in more than one data storage location (for example, more than one server) from within a single application.

- ODBC allows users to access data in more than one type of DBMS (such as IBM DB2, Oracle, Microsoft SQL Server and DEC Rdb) from within a single application.

- ODBC greatly simplifies application development. It is now easier for developers to provide access to data in multiple, concurrent DBMSs.

- ODBC is a portable application programming interface (API), enabling the same interface and access technology to be a cross-platform tool.

- ODBC insulates applications from changes to underlying network and DBMS versions. Modifications to networking transports, servers, and DBMSs will not affect current ODBC applications.

- ODBC promotes the use of SQL—the standard language for DBMSs. It is an open, vendor-neutral specification based on the SAG Call Level Interface (CLI).

Heterogeneous Database Environments

Historically, database applications have been built to access a single source of data. The range of applications varies from mainframe-based, batch-oriented DBMSs, to terminal-based, interactive applications, to personal computer–based, single-user DBMSs, to the more recent client/server DBMSs. Data typically resides in a variety of file formats, such as VSAM and ISAM, as well as in hierarchical and relational DBMSs.

Corporations typically have applications and data residing on diverse platforms and DBMSs for historical, strategic, and technological reasons.

Corporations often have legacy systems that must be maintained because they contain key corporate data. Corporate mergers often bring together diverse information technologies. Systems were often developed using technology that met a specific requirement, such as an engineering application. Departmental users developed their own workgroup and personal computer databases.

There is a strong requirement for a common method of accessing, managing, and analyzing data. The heterogeneous nature of database environments is a problem corporations are faced with today.

Database Connectivity Components

Database connectivity allows an application to communicate with one or more DBMSs. Database connectivity is a requirement whether the application uses a file-based (ISAM) approach, a client/server model, or traditional mainframe connectivity. The requirement for database connectivity has been hastened by client/server computing.

As illustrated in Figure 22-1, some of the key components involved with database connectivity are:

- **Application:** Allows users to perform a set of functions, such as queries, data entry, and report generation. This would be the order status application in our computer telephony (CT) system example.

- **Client system:** The physical system where the client portion of an application runs. In our example, this is the PC that runs the CT application.

- **Data access software:** A service layer on client systems that provides a direct interface for applications. This middleware or enabling software plays a key role in client/server data access. This layer accepts data retrieval and update requests directly from the application, and transmits them across the network. This middleware also is responsible for returning results and error codes back to the application.

- **Network software:** The software protocols that allow the client to communicate with the server system.

- **Network:** The physical connection of the client to the server system.

- **Data source:** The data and method of data access. The data may exist in a variety of hierarchical or relational DBMSs, or in a file with a format such as ISAM or VSAM.

- **Server system:** The physical system where the DBMS resides (also known as the host system). For example, the server system could be a PC, a minicomputer, or a mainframe.

Components of Database Connectivity

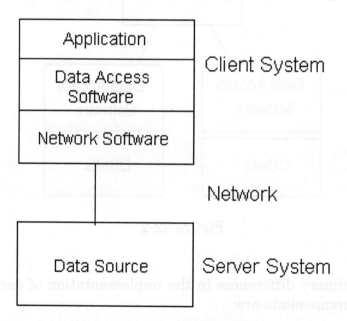

Figure 22-1

The Challenge of Database Connectivity

The problems of database connectivity are apparent in the differences among the programming interfaces, DBMS protocols, DBMS languages, and network protocols of disparate data sources. Even when data sources are restricted to relational DBMSs that use SQL, significant differences in SQL syntax and semantics must be resolved. See figure 22-2.

Multiple Database Connectivity

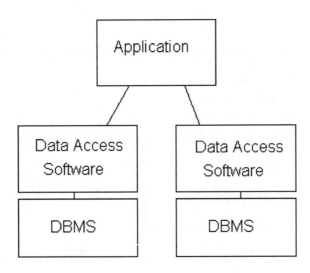

Figure 22-2

The primary differences in the implementation of each of these components are:

- **Programming interface:** Each vendor provides their own proprietary programming interface. One method of accessing a relational DBMS is through embedded SQL. Another method is through an API.

- **DBMS protocol:** Each vendor uses proprietary data formats and methods of communication between the application and the DBMS. For example, there are many different ways to delineate the end of one row of data and beginning of the next.

- **DBMS language:** SQL has become the language of choice for relational DBMSs, but many differences still exist among actual SQL implementations.

- **Networking protocols:** There are many diverse local area network (LAN) and wide area network (WAN) protocols in networks today. DBMSs and applications must coexist in these diverse environments. For example, SQL Server may use DECnet on a VAX, TCP/IP on UNIX, and Netbeui or SPX/IPX on a PC.

Approaches to Database Connectivity

Several vendors have attempted to address the problem of database connectivity in a variety of ways. The primary approaches include using gateways, a common programming interface, and a common protocol.

- **Gateways:** Application developers tend to use one vendor's programming interface, SQL grammar, and DBMS protocol. A gateway causes a target DBMS to appear to the application as a copy of the selected DBMS. The gateway translates and forwards requests to the target DBMS and receives results from it. It is important to note that an application using a gateway would need a different gateway for each type of DBMS it needs to access, such as DEC Rdb, Informix, Ingres, and Oracle. The gateway approach is limited by architectural differences among DBMSs, such as differences in catalogs and SQL implementations, and the need for one gateway for each target DBMS.

- **Common Interface:** With the common interface approach, a single programming interface is provided to the developer. As shown in figure 22-3, it is possible to provide some standardization in a database application development environment or user interface even when the underlying interfaces are different for each DBMS. This is accomplished by creating a standard API, macro language, or set of user tools for accessing data and translating requests for, and results from, each target DBMS. A common interface is usually implemented by writing a driver for each target DBMS.

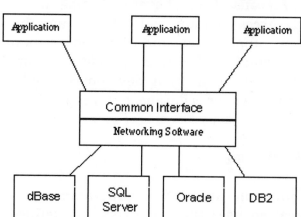

Figure 22-3

- **Common Protocol:** The DBMS protocol, SQL grammar, and networking protocols are common to all DBMSs, so the application can use the same protocol and SQL grammar to communicate with all DBMSs. Examples are remote data access (RDA) and distributed relational database architecture (DRDA). RDA is an emerging standard from SAG, but not available today. DRDA is IBM's alternative DBMS protocol. Common protocols can ultimately work very effectively in conjunction with a common interface.

The ODBC Solution

ODBC addresses the heterogeneous database connectivity problem using the common interface approach.

307

Application developers use one API to access all data sources. ODBC is based on a CLI specification, which was developed by a consortium of over 40 companies (members of the SQL Access Group and others), and has broad support from application and database vendors. The result is a single API that provides most of the functionality that application developers need, and an architecture that ensures database interoperability.

How ODBC Works

ODBC defines an application programming interface. Each application uses the same code, as defined by the API specification, to talk to many types of data sources through DBMS-specific drivers. A Driver Manager sits between the applications and the drivers. In Windows, the Driver Manager and the drivers are implemented as *dynamic link libraries* (DLLs).

The application calls ODBC functions to connect to a data source, send and receive data, and disconnect.

The Driver Manager provides information to an application such as a list of available data sources; loads drivers dynamically as they are needed; and provides argument and state transition checking.

The driver, developed separately from the application, sits between the application and the network. The driver processes ODBC function calls, manages all exchanges between an application and a specific DBMS, and may translate the standard SQL syntax into the native SQL of the target data source. All SQL translations are the responsibility of the driver developer.

Applications are not limited to communicating through one driver. A single application can make multiple connections, each through a different driver, or multiple connections to similar sources through a single driver.

To access a new DBMS, a user or an administrator installs a driver for the DBMS. The user does not need a different version of the application to access the new DBMS.

What ODBC Means to Application Developers

ODBC was designed to allow application developers to decide between using the least common denominator of functionality across DBMSs or exploiting the individual capabilities of specific DBMSs.

ODBC defines a standard SQL grammar and set of function calls that are based upon the SAG CLI specification, called the *core grammar* and *core functions*, respectively. If an application developer chooses only to use the core functionality, they need not write any additional code to check for specific capabilities of a driver.

With core functionality, an application can:

- Establish a connection with a data source, execute SQL statements, and retrieve results.

- Receive standard error messages.

- Provide a standard log-on interface to the end user.

- Use a standard set of data types defined by ODBC.

- Use a standard SQL grammar defined by ODBC.

ODBC also defines an extended SQL grammar and a set of extended functions to provide application developers with a standard way to exploit advanced capabilities of a DBMS. In addition to the above features, ODBC includes a set of extensions that provide enhanced performance and increased power through the following features:

- Data types such as date, time, timestamp, and binary.

- Scrollable cursors.

- A standard SQL grammar for scalar functions, outer joins, and procedures.

- Asynchronous execution.

- A standard way for application developers to find out what capabilities a driver and data source provide.

Finally, ODBC supports the use of DBMS-specific SQL grammar, allowing applications to exploit the capabilities of a particular DBMS.

Industry Commitment to ODBC

The following is a partial list of databases currently supported by one or more ODBC database drivers:

Access	Gupta SQLBase	Paradox
Btrieve	Informix	Progress
DB2	Ingres	Rdb
dBASE	NetWareSQL	SQL Server
Excel	NonStop SQL	UNIFY
FoxPro	Oracle	WATCOM SQL

ODBC Architecture

Figure 22-4 illustrates the ODBC architecture:

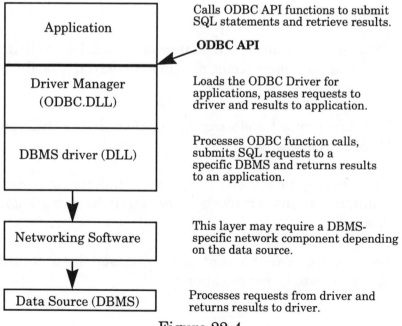

Figure 22-4

311

ODBC Interface

The ODBC interface defines the following:

- A library of ODBC function calls that allow an application to connect to a DBMS, execute SQL statements, and retrieve results.

- SQL syntax based on the X/Open and SQL Access Group (SAG) SQL CAE specification (1992).

- A standard set of error codes.

- A standard way to connect and log-on to a DBMS.

- A standard representation for data types.

The interface is flexible:

- Strings containing SQL statements can be explicitly included in source code or constructed on the fly at run-time.

- The same object code can be used to access different DBMS products.

- An application can ignore underlying data communications protocols between it and a DBMS product.

- Data values can be sent and retrieved in a format convenient to the application.

The ODBC interface provides two types of function calls:

- Core functions are based on the X/Open and SQL Access Group Call Level Interface specification.

- Extended functions support additional functionality, including scrollable cursors and asynchronous processing.

To send an SQL statement, the statement is included as an argument in an ODBC function call. The statement need not be customized for a specific DBMS.

ODBC Components

The ODBC architecture has four components:

- **Application:** Performs processing and calls ODBC functions to submit SQL statements and retrieve results.

- **Driver Manager**: Loads drivers on behalf of an application.

- **Driver:** Processes ODBC function calls, submits SQL requests to a specific data source, and returns results to the application. If necessary, the driver modifies an application's request so that the request conforms to syntax supported by the associated DBMS.

- **Data Source:** Consists of the data the application wants to access and its associated operating system,

DBMS, and network platform (if any) used to access the DBMS.

Application

An application using the ODBC interface performs the following tasks:

- Requests a connection, or session, with a data source.

- Sends SQL requests to the data source.

- Defines storage areas and data formats for the results of SQL requests.

- Requests results.

- Processes errors.

- Reports results back to a user, if necessary.

- Requests commit or rollback operations for transaction control.

- Terminates the connection to the data source.

An application can provide a variety of features external to the ODBC interface, including online transaction processing, and report generation; the application may or may not interact with users. In our example, computer telephony order status inquiry is the application

Driver Manager

The Driver Manager, provided by Microsoft, is a dynamic-link library (DLL) with an import library. The primary purpose of the Driver Manager is to load drivers. The Driver Manager and driver appear to an application as one unit that processes ODBC function calls. The Driver Manager also performs the following:

- Uses the ODBC.INI file or registry to map a data source name to a specific driver dynamic-link library (DLL).

- Processes several ODBC initialization calls.

- Provides entry points to ODBC functions for each driver.

- Provides parameter validation and sequence validation for ODBC calls.

Driver

A driver is a DLL that implements ODBC function calls and interacts with a data source. The Driver Manager loads a driver when the application calls the SQLBrowseConnect, SQLConnect, or SQLDriverConnect function.

A driver performs the following tasks in response to ODBC function calls from an application:

- Establishes a connection to a data source.

315

- Submits requests to the data source.

- Translates data to or from other formats, if requested by the application.

- Returns results to the application.

- Formats errors into standard error codes and returns them to the application.

- Declares and manipulates cursors if necessary. (This operation is invisible to the application unless there is a request for access to a cursor name.)

- Initiates transactions if the data source requires explicit transaction initiation. (This operation is invisible to the application.)

Data Source

DBMS refers to the general features and functionality provided by an SQL database management system. A data source is a specific instance of a combination of a DBMS product and any remote operating system and network necessary to access it.

An application establishes a connection with a particular vendor's DBMS product on a particular operating system, accessible by a particular network. For example, the application might establish connections to:

- An Oracle DBMS running on an OS/2 operating system, accessed by Novell netware.

- A local Xbase file, in which case the network and remote operating system are not part of the communication path.

- A Tandem NonStop SQL DBMS running on the Guardian 90 operating system, accessed via a gateway.

Types of Drivers

ODBC defines two types of drivers:

- **Single-tier**: The driver processes both ODBC calls and SQL statements. (In this case, the driver performs part of the data source functionality.)

- **Multiple-tier**: The driver processes ODBC calls and passes SQL statements to the data source.

One system can contain both types of configurations.

Single-Tier Configuration

In a single-tier implementation, the database file is processed directly by the driver. The driver processes SQL statements and retrieves information from the database. A driver that manipulates an Xbase file is an example of a single-tier implementation.

A single-tier driver may limit the set of SQL statements that may be submitted. The minimum set of SQL statements that must be supported by a single-tier driver is defined in "SQL Grammar." Figure 22-5 shows two types of single tier configurations.

Single Tier Implementations

System A

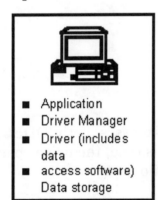

Client B   **Server B**

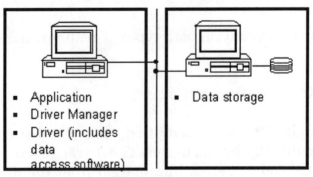

Figure 22-5

Multiple-Tier Configuration

In a multiple-tier configuration, the driver sends SQL requests to a server that processes SQL requests.

Although the entire installation may reside on a single system, it is more often divided across platforms. The application, driver, and Driver Manager reside on one system, called the client. The database and the software that controls access to the database typically reside on another system, called the server.

Another type of multiple-tier configuration is a gateway architecture. The driver passes SQL requests to a gateway process, which in turn sends the requests to the data source.

Figure 22-6 shows three types of multiple-tier configurations. From an application's perspective, all three configurations are identical.

Matching an Application to a Driver

One of the strengths of the ODBC interface is interoperability; a programmer can create an ODBC application without targeting a specific data source.

Users can add drivers to the application after it is compiled and shipped.

Multiple Tier Implementations

System C

- Application
- Driver Manager
- Driver
- Data access
- software
  Data storage

Client D   **Server D**

- Application   - Data access
- Driver Manager   - software
- Driver   Data storage

Client E   **Server E1**   **Server E2**

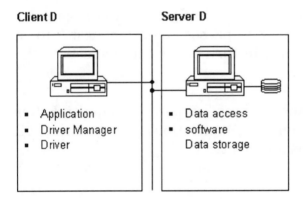

- Application   - Gateway software   - Data access
- Driver Manager   - software
- Driver   Data storage

Figure 22-6

From an application standpoint, it would be ideal if every driver and data source supported the same set of ODBC function calls and SQL statements. However, data sources and their associated drivers provide a varying range of functionality. Therefore, the ODBC interface defines conformance levels, which determine the ODBC procedures and SQL statements supported by a driver.

ODBC Conformance Levels

ODBC defines conformance levels for drivers in two areas:

- the ODBC API; and,

- the ODBC SQL grammar (which includes the ODBC SQL data types).

Conformance levels help both application and driver developers by establishing standard sets of functionality. Applications can easily determine if a driver provides the functionality they need. Drivers can be developed to support a broad selection of applications without being concerned about the specific requirements of each application.

To claim that it conforms to a given API or SQL conformance level, a driver must support all the functionality in that conformance level, regardless of whether that functionality is supported by the DBMS associated with the driver. However, conformance levels do not restrict drivers to the functionality in the levels to which they conform. Applications can determine the functionality supported by a driver by calling SQLGetInfo, SQLGetFunctions, and SQLGetTypeInfo.

API Conformance Levels

The ODBC API defines a set of core functions that correspond to the functions in the X/Open and SQL Access Group Call Level Interface specification. ODBC also defines two extended sets of functionality, Level 1 and Level 2. The following list summarizes the functionality included in each conformance level. Many ODBC applications require that drivers support all of the functions in the Level 1 API conformance level.

Core API

- Allocate and free environment, connection, and statement handles.

- Connect to data sources. Use multiple statements on a connection.

- Prepare and execute SQL statements. Execute SQL statements immediately.

- Assign storage for parameters in an SQL statement and result columns.

- Retrieve data from a result set. Retrieve information about a result set.

- Commit or roll back transactions.

- Retrieve error information.

Level 1 API

- Core API functionality.

- Connect to data sources with driver-specific dialog boxes.

- Set and inquire values of statement and connection options.

- Send part or all of a parameter value (useful for long data).

- Retrieve part or all of a result column value (useful for long data).

- Retrieve catalog information (columns, special columns, statistics, and tables).

- Retrieve information about driver and data source capabilities, such as supported data types, scalar functions, and ODBC functions.

Level 2 API

- Core and Level 1 API functionality.

- Browse connection information and list available data sources.

- Send arrays of parameter values. Retrieve arrays of result column values.

- Retrieve the number of parameters and describe individual parameters.

- Use a scrollable cursor.

- Retrieve the native form of an SQL statement.

- Retrieve catalog information (privileges, keys, and procedures).

- Call a translation DLL.

SQL Conformance Levels

ODBC defines a core grammar that roughly corresponds to the X/Open and SQL Access Group SQL CAE specification (1992). ODBC also defines a minimum grammar, to meet a basic level of ODBC conformance, and an extended grammar, to provide for common DBMS extensions to SQL. The following list summarizes the grammar included in each conformance level.

Minimum SQL Grammar

- Data Definition Language (DDL): CREATE TABLE and DROP TABLE.

- Data Manipulation Language (DML): simple SELECT, INSERT, UPDATE SEARCHED, and DELETE SEARCHED.

- Expressions: simple (such as A > B + C).

- Data types: CHAR, VARCHAR, or LONG VARCHAR.

Core SQL Grammar

- Minimum SQL grammar and data types.

- DDL: ALTER TABLE, CREATE INDEX, DROP INDEX, CREATE VIEW, DROP VIEW, GRANT, and REVOKE.

- DML: full SELECT.

- Expressions: subquery, set functions such as SUM and MIN.

- Data types: DECIMAL, NUMERIC, SMALLINT, INTEGER, REAL, FLOAT, DOUBLE PRECISION.

Extended SQL Grammar

- Minimum and Core SQL grammar and data types.

- DML: outer joins, positioned UPDATE, positioned DELETE, SELECT FOR UPDATE, and unions.

- Expressions: scalar functions such as SUBSTRING and ABS, date, time, and timestamp literals.

- Data types: BIT, TINYINT, BIGINT, BINARY, VARBINARY, LONG VARBINARY, DATE, TIME, TIMESTAMP

- Batch SQL statements.

- Procedure calls.

How to Select a Set of Functionality

The ODBC functions and SQL statements that a driver supports usually depend on the capabilities of its associated data source. The ODBC functions and SQL statements that an application uses depend on:

- The functionality needed by the application.

- The performance needed by the application.

- The data sources to be accessed by the application and the extent to which the application must be inter-operable among these data sources.

- The functionality available in the drivers used by the application.

Because drivers support different levels of functionality, application developers may have to make trade-offs among the factors listed above.

Developers of specialized applications may make different trade-offs than developers of generalized applications. For example, the developer of an application that only transfers data between two DBMSs (each from a different vendor) can safely exploit the full functionality of each of the drivers.

Connections and Transactions

Before an application can use ODBC, it must initialize ODBC and request an environment handle (henv). To communicate with a data source, the application must request a connection handle (hdbc) and connect to the data source. The application uses the environment and connection handles in subsequent ODBC calls to refer to the environment and specific connection.

An application may request multiple connections for one or more data sources. Each connection is considered a separate transaction space.

An active connection can have one or more statement processing streams.

A driver maintains a transaction for each active connection. The application can request that each SQL statement be automatically committed on completion; otherwise, the driver waits for an explicit commit or rollback request from the application. When the driver performs a commit or rollback operation, the driver resets all statement requests associated with the connection.

The Driver Manager allows an application to switch connections while transactions are in progress on the current connection.

Designing CT Applications with ODBC

The Need for Performance in CT Applications

Two factors combine to demand that Computer Telephony applications be designed for high performance:

- Simultaneous handling of multiple telephone lines

- Caller aversion to "dead air"

Most ODBC applications are written with the assumption that a client will have a single connection to the database server. However a CT system can easily have one or more connection for each phone line processed. Therefore a 24 line T-1 based system could appear to the server as a client with 24 ODBC connections making requests at a rate unheard of for a traditional workstation.

In addition, after about two seconds of silence on the line, callers get uneasy and start thinking about hanging up. This means that a response time that is acceptable for a workstation operator can be unacceptable for a caller to a CT system.

These two factors conspire to make multi-line computer telephony applications one of the most demanding environments for ODBC implementations.

The Myth: ODBC Is Slow

Some technical professionals believe that ODBC performance is poor. The net effect of this perception is that this group of people position ODBC as a low performance solution for decision support applications to be used only when connectivity to multiple databases is an absolute requirement.

Most of their performance issues can be attributed to one or more of the following:

- Inappropriate selection of development tools for the desired level of functionality

- Lack of understanding or under-utilization of the database technology

- Un-optimized ODBC drivers

Of course there are situations where you should expect decreased performance from ODBC. For example, ODBC access to non-SQL databases such as Xbase (Clipper, dBASE and FoxPro) will be slower in certain cases than going directly through Xbase. PC databases are built on the ISAM (Indexed Sequential Access Method) model which is different that the SQL model. Some things are much easier with ISAMs, such as scrolling, single table operations and indexed operations. However others are more difficult, if not impossible, such as joins, aggregations, expressions and general Boolean queries. Neither model is better; it depends on the features required by the application.

Comparative Performance Tests: ODBC vs. Native

Resource Group, Inc. recently conducted verifiable, independent tests comparing performance of ODBC Drivers versus proprietary SQL programming interfaces for the following database systems:

- Oracle 7

- INFORMIX 5.01

- Sybase System 10.0.2

The tests used a variety of client and server hardware and software and included logic that uses various SQL programming techniques.

An examination of the mean execution time of 29 tests revealed that 12 tests were faster with native performance while 17 were faster with ODBC performance. In the 17 tests where ODBC had a performance advantage, the difference was greater than 10% in 8 test. For the 12 tests where native performance was better, the difference was greater than 10% in only 2 tests. These test demonstrate that performance when using ODBC may often be superior to native performance; and that when native performance is superior, the difference is often minor.

High Performance ODBC Drivers

ODBC drivers are the critical link between ODBC-compliant applications and data sources. However, all

ODBC drivers are not created equal. Many ODBC drivers on the market (especially the "free" drivers) are not optimized. That is the polite way of saying they are performance hogs.

INTERSOLV, Inc. is the leading provider of database access technology to major software publishers, including Microsoft, IBM, Informix, Sybase and Oracle. Their DataDirect ODBC Drivers provide the industry's most comprehensive range of ODBC technology, providing access to over 35 databases and formats.

INTERSOLV DataDirect Driver highlights include:

• Supports all major SQL and PC databases

• State-of-the art technology conforming to the Microsoft ODBC 2.5 specification

• Consistent level of ODBC Core, Level 1 and Level 2 functions

• Optimal reliability and performance

Designing High Performance ODBC Applications

Designing and coding performance-oriented applications is not easy. There is no "Performance" chapter in the *ODBC Programmer's Reference*. There are no warnings returned from ODBC drivers or the ODBC driver manager when applications are coded inefficiently. Furthermore, there are no internal documents describing how ODBC drivers are implemented.

The following section provides guidelines for designing performance-oriented ODBC applications. These guidelines were accumulated by the bright people at INTERSOLV, Inc. by examining the implementation of several dozen released ODBC applications. Perhaps the "ODBC is slow" myth has grown from the release of some of these un-optimized applications.

The guidelines presented are not hard theorems but general advice. While some ODBC drivers may not benefit from one particular guideline, these are the guidelines that benefit the majority of drivers. The guidelines are based on four general rules:

Problem	Solution
Network communication is slow	Reduce network traffic
Complex queries are slow and reduce concurrency	Simplify queries
Excessive calls decrease performance	Optimize application to driver interaction
Disk I/O is slow	Limit disk I/O

Managing Connections

Connection management is extremely important to application performance. Designers should optimize applications by connecting once and using multiple statement handles instead of performing multiple connections.

Connection management is one of the most poorly designed elements of existing ODBC applications. Connecting to a data source is extremely expensive and should be avoided after establishing an initial connection.

Some ODBC applications are designed such that they call informational gathering routines that have no record of already attached connection handles. For example, some applications establish a connection and then call a routine in a separate DLL or shared library that re-attaches and gathers up-front information about the driver to be used later in the application. While gathering driver information at connect time is a good algorithm, it should not be minimized by connecting twice to get this information. At least one popular ODBC enabled application connects a second time to gather driver information but never disconnects the second connection. Applications that are designed as separate entities should pass the already connected HDBC pointer to the data collection routine instead of establishing a second connection.

Another similarly poor algorithm is to connect and disconnect several times throughout your application to perform SQL statements. Connection handles can have multiple statement handles associated with them. Statement handles are defined to be memory storage for information about SQL statements. Why then do many applications allocate new connection handles to perform SQL statements? Applications should use statement handles to manage multiple SQL statements.

Connection and statement handling should not be delayed until implementation. Spending time in the design phase on connection management makes your application perform better and most certainly makes it more maintainable.

Committing Data

Committing data is extremely disk i/o intensive and thus slow. Always turn auto-commit off if the driver can support transactions.

What is actually involved in a commit? At commit time the DBMS server must flush back to disk every data page that contains updated or new data. Note that this is not a sequential write but a searched write to replace existing data already in the table. By default, auto-commit is on when connecting to a data source. Auto-commit mode is typically detrimental to performance because of the extreme amount of disk i/o needed to commit every operation.

Further reducing performance, some DBMS servers do not provide an "auto-commit mode". For this type of server, the ODBC driver must explicitly issue a COMMIT statement and perhaps a BEGIN TRANSACTION for every operation sent to the server. In addition to the large amount of disk i/o required to support auto-commit mode, we must also pay a performance penalty for up to three network requests for every statement issued by an application in this scenario.

Asynchronous Execution

Design your application to take advantage of data sources that support asynchronous execution. Asynchronous calls do not perform faster but well designed applications appear to run more efficiently.

By default, an application makes calls to an ODBC driver that then executes statements against the DBMS server in a synchronous manner. In this mode of operation the driver does not return control to the application until its own request to the server is complete. For statements which take more than a few seconds to complete execution, this can result in the perception of poor performance to the end user.

Some data sources support asynchronous execution. When in asynchronous mode, an application makes calls to an ODBC driver, and control is returned almost immediately. In this mode the driver returns the status SQL_STILL_EXECUTING to the application and then sends the appropriate request to the database back-end for execution.

The application polls the driver at various intervals at which point the driver itself polls the server to see if the query has completed execution. If the query is still executing, then the status SQL_STILL_EXECUTING is returned to the application. If it has completed, then a status such as SQL_SUCCESS is returned, and the application can then begin to fetch records.

Turning on asynchronous execution does not by itself improve performance. Well designed applications, however, can take advantage of asynchronous query execution by allowing the end user to work on other things while the query is being evaluated on the server. Perhaps users will start one or more subsequent queries or choose to work in another application, all while the query is executing on the server. Designing for asynchronous execution makes your application appear to run faster by allowing the end user to work concurrently on multiple tasks.

Catalog Functions

Catalog functions such as SQLColumns and SQLTables are relatively slow compared to all other ODBC functions. Applications should cache information returned from catalog functions so that multiple executions are not needed.

While almost no ODBC application can be written without catalog functions, their use should be minimized. To return all result column information mandated by the ODBC specification, a driver may have to perform multiple queries, joins, subqueries, and/or unions in order to return the necessary result set for a single call to a catalog function. These particular elements of the SQL language are performance hogs. Frequent use of catalog functions in an application will likely result in poor performance.

Applications should attempt to cache information from catalog functions if possible. For example, call SQLGetTypeInfo once in the application and store away the elements of the result set that your application depends on. It is unlikely that any application uses all elements of the result set generated by a catalog function so the cache of information should not be hard to maintain.

Passing Null Arguments

Passing null arguments to catalog functions results in time consuming queries being generated by the driver. In addition, network traffic potentially increases due to unwanted result set information. Always supply as many non-null arguments to catalog functions as possible.

In some circumstances not much information is known about the object for which you are requesting information, but in many cases at least some information is known. Any additional information that the application can send the driver when calling a catalog function can result in improved performance and reliability.

SQLColumns

Avoid using SQLColumns to determine characteristics about a table. Instead, use a dummy query with SQLDescribeCol. Consider an ad-hoc application that allows the user to choose the columns that will be selected.

Should the application use SQLColumns to return information about the columns to the user or instead prepare a dummy query and call SQLDescribeCol?

In both cases a query is sent to the server, but in Case 1 the query must be evaluated and form a result set that must be sent to the client.

Clearly, Case 2 is the better performing model.

Retrieving Long Data

Retrieving long data (SQL_LONGVARCHAR and SQL_LONGVARBINARY data) across the network is very resource intensive and thus slow. Applications should avoid requesting long data unless it is absolutely necessary.

How often do users want long data? By default, it is generally acceptable not to retrieve long data or binary data because most users don't want such information. If the user does want these result items, then the application can re-query the database specifying only the long columns in the select list. This method allows the average user to retrieve the result set without having to pay a high performance penalty for intense network traffic.

While the most optimal method is to exclude long data from the select list, some applications do not formulate the select list before sending the query to the ODBC driver (i.e., some applications simply 'select * from <table name>...'). If the select list contains long data then some drivers must retrieve that data at fetch time even if the application does not bind the long data in the result set. If possible, the designer should attempt to implement a method that does not retrieve all columns of the table.

Reducing the Size of Data Retrieved

Reduce the size of any data being retrieved to some manageable limit by calling SQLSetStmtOption with the SQL_MAX_SIZE option. This reduces network traffic and improves performance.

While eliminating SQL_LONGVARCHAR and SQL_LONGVARBINARY data from the result set is ideal in terms of performance, many times long data must be retrieved. Consider, however, that most users do not want 100k or more of textual information. What techniques, if any, are available to limit the amount of data retrieved?

Many application developers mistakenly assume that if they call SQLGetData with a container of size x that the ODBC driver only retrieves x bytes of information from the server. Because SQLGetData can be called multiple times for any one column, most drivers optimize their network use by retrieving long data in large chunks and then returning it to the user when requested.

One 64k retrieval is less expensive than sixty-four 1000 byte retrievals in terms of network access. Unfortunately, the application may not call SQLGetData again; thus, the first and only retrieval is slowed by the fact that 64k of data has to be sent across the network.

Many ODBC drivers allow limiting the amount of data retrieved across the network by supporting the statement option SQL_MAX_SIZE. This option allows the driver to communicate to the DBMS back-end server that only z bytes of data are pertinent to the client. The server responds by sending only the first z bytes of data for all result columns. This optimization greatly reduces network traffic and improves performance of the client.

Using Bound Columns

Retrieving data through bound columns (SQLBindCol) instead of using SQLGetData reduces the ODBC call load

339

and thus improves performance. In addition to reducing the call load many drivers optimize use of SQLBindCol by binding result information directly from the DBMS into the user's buffer. That is, instead of the driver retrieving information into a container then copying that information to the user's buffer, the driver simply requests the information from the server be placed directly into the user's buffer.

Using SQLExtendedFetch instead of SQLFetch

Use SQLExtendedFetch to retrieve data instead of SQLFetch. The ODBC call load decreases (resulting in better performance), and the code is less complex (resulting in more maintainable code).

Most ODBC drivers now support SQLExtendedFetch for forward only cursors; yet, most ODBC applications use SQLFetch to retrieve data. In addition to reducing the call load, many ODBC drivers retrieve data from the server in arrays that further improves the performance by reducing network traffic.

For those drivers that do not support SQLExtendedFetch, the application can enable forward only cursors using the ODBC cursor library (call SQLSetConnectOption using SQL_ODBC_CURSORS/SQL_CUR_USE_IF_NEEDED). While using the cursor library does not improve performance, it should not be detrimental to application response time when using forward only cursors (no logging is required). Furthermore, using the cursor library when SQLExtendedFetch is not supported natively by the driver simplifies the code because the application can always depend on SQLExtendedFetch being available. The application need not code two algorithms (one using SQLExtendedFetch and one using SQLFetch).

SQLPrepare/SQLExecute Vs. SQLExecDirect

Don't assume that SQLPrepare/SQLExecute is always as efficient as SQLExecDirect. Use SQLExecDirect for queries that will be executed once and SQLPrepare/ SQLExecute for queries that will be executed more than once.

ODBC drivers are optimized based on the perceived use of the functions that are being executed. SQLPrepare/ SQLExecute is optimized for multiple executions of a statement that most likely uses parameter markers. SQLExecDirect is optimized for a single execution of an SQL statement. Unfortunately, more than seventy-five percent of all ODBC applications use SQLPrepare/ SQLExecute exclusively.

The pitfall of always coding SQLPrepare/SQLExecute can be understood better by considering an ODBC driver that implements SQLPrepare by creating a stored procedure on the server that contains the prepared statement. Creating a stored procedure has substantial overhead, but what the ODBC driver is assuming is that the statement will be executed multiple times. While stored procedure creation is relatively expensive, execution is minimal because the query is parsed and optimization paths are stored at create procedure time. Using SQLPrepare/ SQLExecute for a statement that will be executed only once with such an ODBC driver will result in unneeded overhead. Furthermore, applications that use

SQLPrepare/SQLExecute for large single execution query batches will almost certainly exhibit poor performance when used with ODBC drivers as previously discussed.

Similar arguments can be used to show applications that always use SQLExecDirect cannot perform as well as those that logically use a combination of SQLPrepare/ SQLExecute and SQLExecDirect sequences.

Using SQLPrepare and Multiple SQLExecute Calls

Applications that use SQLPrepare and multiple SQLExecute calls should use SQLParamOptions if available. Passing arrays of parameter values reduces the ODBC call load and greatly reduces network traffic.

Some ODBC drivers do not support SQLParamOptions but many drivers do. To achieve high performance, applications should contain algorithms for using SQLParamOptions if the ODBC driver supports the function. SQLParamOptions is ideal for copying data into new tables or bulk loading tables.

Using the Cursor Library

Do not automatically use the cursor library if scrollable cursors are provided by the driver. The cursor library creates local temporary log files, which are expensive to generate and provide worse performance than using native scrollable cursors.

The cursor library adds support for static cursors, which simplifies the coding of applications that use scrollable cursors. However, the cursor library creates temporary log files on the user's local disk drive to accomplish the task. Disk i/o is typically one of the slowest operations on personal computers. While the benefits of the cursor library is great, applications should not automatically choose to use the cursor library if an ODBC driver supports scrollable cursors natively.

ODBC drivers that support scrollable cursors typically achieve high performance by requesting that the DBMS server produce a scrollable result set instead of emulating the capability by creating log files.

Using Positional Updates and Deletes

Use positional updates and deletes or SQLSetPos whenever possible to update data.

Designing an efficient method for updating data is difficult. While positioned updates do not apply to all types of applications, developers should attempt to use positioned updates and deletes whenever possible. Positioned updates (either via "update where current of cursor" or via SQLSetPos) allow the developer to update data simply by positioning the database cursor to the appropriate row to be changed and signaling the driver to "change the data here". The designer is not forced to build a complex SQL statement but is simply required to supply the data that is to be changed.

Besides making the code more maintainable, positioned updates typically result in improved performance. Because the database server is already positioned on the row (for the Select statement currently in process),

343

expensive operations to locate the row to be changed are not needed. If the row must be located then the server typically has an internal pointer to the row available (for example, ROWID).

Using SQLSpecialColumns

Use SQLSpecialColumns to determine the most optimal set of columns to use in the Where clause for updating data. Many times pseudo-columns provide the fastest access to the data, and these columns can only be determined by using SQLSpecialColumns. Many applications cannot be designed to take advantage of positional updates and deletes. These applications typically update data by forming a Where clause consisting of some subset of the column values returned in the result set. Some applications may formulate the Where clause by using all searchable result columns or by calling SQLStatistics to find columns that may be part of a unique index. These methods typically work but may result in fairly complex queries.

Applications should call SQLSpecialColumns/ SQL_BEST_ROWID to retrieve the most optimal set of columns (possibly a pseudo-column) that identifies any given record. Many databases support special columns that are not explicitly defined by the user in the table definition but are "hidden" columns of every table (for example, ROWID, TID, etc.). These pseudo-columns almost always provide the fastest access to the data because they typically are pointers to the exact location of the record. Since pseudo-columns are not part of the explicit table definition they are not returned from

SQLColumns. The only method of determining if pseudo-columns exist is to call SQLSpecialColumns.

If your data source does not contain special pseudo-columns, then the result set of SQLSpecialColumns consists of the columns of the most optimal unique index on the specified table (if a unique index exists); therefore, your application need not additionally call SQLStatistics to find the smallest unique index.

EASE for Windows NT ODBC Server and Query Builder

What is the EASE ODBC Server?

The EASE ODBC Server is a separate, independent process running under Windows NT in a client-server architecture. The EASE ODBC Server manages pools of persistent connections to multiple local and remote ODBC compliant data sources. All EASE ODBC data transactions are handled by the ODBC Server. And these data transactions are processed asynchronously so that the cursor for a quick query on phone line two can be returned to the application before the cursor for a complex query initiated earlier on phone line one.

Typical Operation of the EASE ODBC Server

Configuration:
1. Set global properties for both connections and cursors. Global properties will apply to all connections and cursors

2. Define each type of connection:

   a. Alias name and corresponding ODBC data source name

   b. Number of persistent connections to be created to data source (pool size)

   c. Override default connection properties as desired

3. Create specified number of persistent connections to ODBC data sources for each connection definition. (connection pools)

<u>Runtime:</u>
4. Get a specific connection from the connection pool

5. Execute a SELECT to create a data cursor

6. Get data from cursor and speak to caller

7. Execute an UPDATE to save any new data

8. Release connection back to the pool

9. Hang-up phone, go to step 4 and wait for next call.

EASE ODBC Server Configuration

All startup configuration parameters are specified in EASE under an ODBC configuration screen (see Figure 22-7). Connection and cursor objects have properties and all properties have default values. The user can override default property values when needed for use with specific ODBC database servers and for performance tuning.

ODBC Configuration Screen

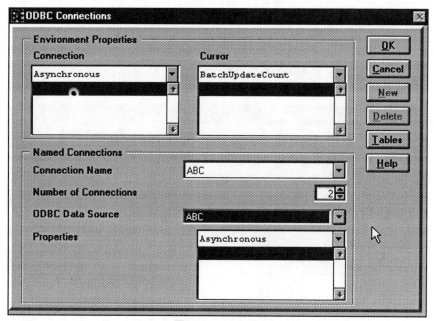

Figure 22-7

Properties can be applied to all (global) or to individual connections and cursors (objects). Global properties are only set during startup, whereas individual properties can be set at startup or during runtime. Specific EASE actions - Set Connection and Set Cursor Property actions allow runtime control of properties. In a typical CT application it is necessary to override only one or two default properties.

ODBC Data Sources

When developing an ODBC application, the first step is to define data sources which provide a symbolic name and location for each group of database files. Figure 22-8 shows the ODBC data source definition named 'ABC' for the sample order status application.

EASE ODBC Driver Setup

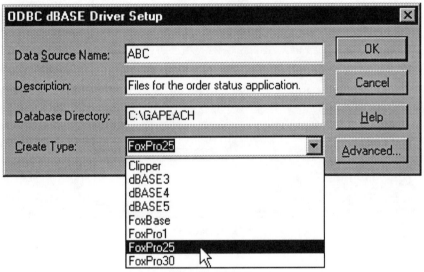

Figure 22-8

Each of the database files is tied to one of the data source names to provide the location of the files. Note the data sources, however, are defined outside of the application environment allowing the files within a data source to be moved to other locations and machines by only updating the data source definition. No application level code needs to be modified.

The structure of the database files required by the application are defined in the EASE ODBC Module. For the Order Status application, the main ORDER database is defined in Figure 22-9.

EASE ODBC Field Import

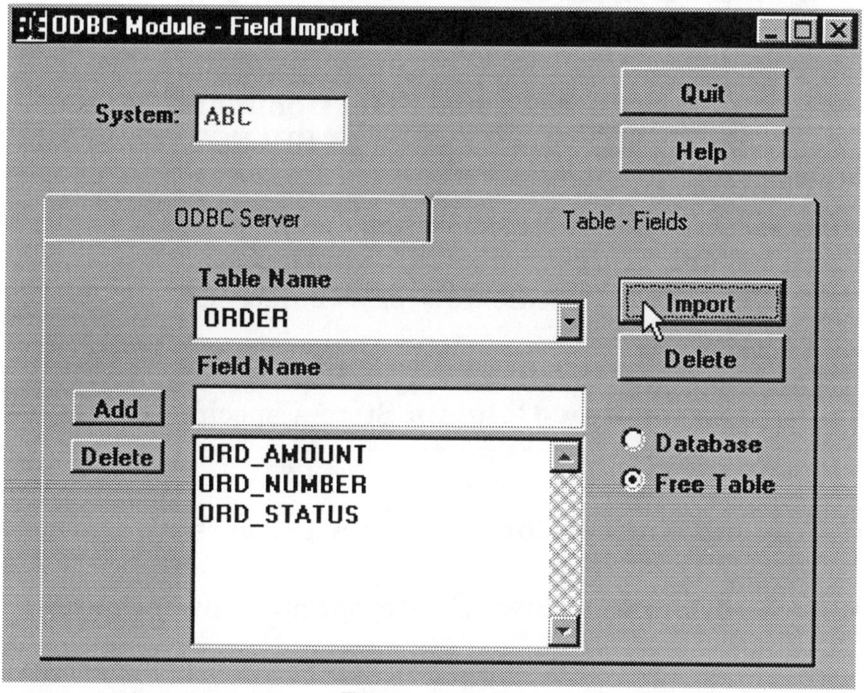

Figure 22-9

Fields can be entered manually by the user or imported automatically. When fields can be imported automatically, the developer is confident that the ODBC configuration and driver setup has been successful. Database tables can be single files or part of a database project.

EASE ODBC Commands

The 16 EASE ODBC Actions include:

- **Connect to Data Source:** Get a connection handle from specified pool

- **Cancel Asynchronous Execution:** Cancel current executing command.

- **Execute Asynchronous SQL Command:** Execute any valid SQL command. e.g. SELECT, INSERT, DELETE, UPDATE.

- **Get/Fetch Cursor Row-Column:** Get one or all columns in a specific row of a cursor created by a SELECT.

- **Put Cursor Row-Column:** Store a specific column of data in a specific cursor row.

- **Append Cursor Row:** Add a row to a cursor.

- **Delete Cursor Row:** Delete specific row in specific cursor.

- **Get Connect Status:** Get current state of current active connection.

- **Disconnect from Data Source:** Release connection handle back to pool.

- **Get Connection Property:** Read a specific connection property.

- **Set Connection Property:** Set a specific connection property to a specific value.

- **Get Cursor Property:** Read a specific cursor property.

- **Set Cursor Property:** Set a specific cursor property to a specific value.

- **Commit Transactions:** Commit all uncommitted transactions.

- **Rollback Transactions:** Rollback transactions to previous commit.

- **Get Last Error Information:** Get detailed information about last ODBC error.

EASE ODBC Query Builder

The EASE ODBC Query Builder is a tool used to quickly and easily create SQL statements using the tables and variables defined in a EASE CT application. Of course, knowledgeable developers can create free-form SQL statements for more complex queries and to accomplish desired performance tuning of an application. This dual mode design provides ease of use for the vast majority of queries while still allowing complete flexibility for the power user.

Figure 22-10 shows the EASE Query Builder in action creating a "Where Clause".

EASE Query Builder - Where Clause

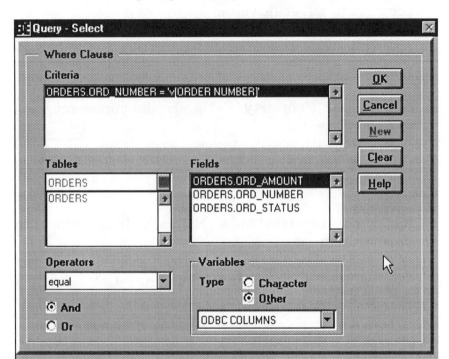

Figure 22-10

The EASE SQL Query Builder is used to define the actions to be performed on the database files by the application.

Figure 22-11 shows the Query Builder screen which defines the completed ORDER INFO query. The ORDER INFO query selects the ORD_AMOUNT and ORD_STATUS fields that contain the amount and status for an order where the contents ORD_NUMBER field equals the EASE application variable ORDER NUMBER.

ODBC Query - Complete

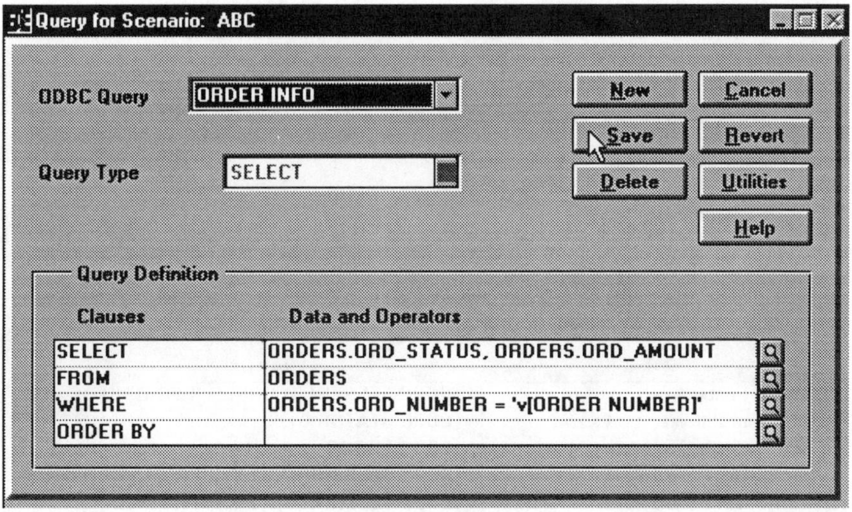

Query for Scenario: ABC		
ODBC Query	ORDER INFO ▼	New Cancel
		Save Revert
Query Type	SELECT ▦	Delete Utilities
		Help

Query Definition

Clauses	Data and Operators	
SELECT	ORDERS.ORD_STATUS, ORDERS.ORD_AMOUNT	🔍
FROM	ORDERS	🔍
WHERE	ORDERS.ORD_NUMBER = 'v[ORDER NUMBER]'	🔍
ORDER BY		🔍

Figure 22-11

The Query Builder can define SQL commands to Update, Insert, Delete and Select records from database files. The output from the Query Builder is the complete SQL statement to access the database.

Finally, the application performs any number of ODBC Execute statements to execute the queries which retrieve or update records from the database files.

In Figure 22-12, the GET ORDER STATUS Label uses the ODBC Execute Action to perform the ORDER INFO Query. The EASE Action ODBC Get Row/Column retrieves the Order Status and Order Amount data from the selected cursor and puts it in the appropriate variables. The data is now available to the CT application for our final example of "Giving Data a Voice".

353

ODBC Data Access

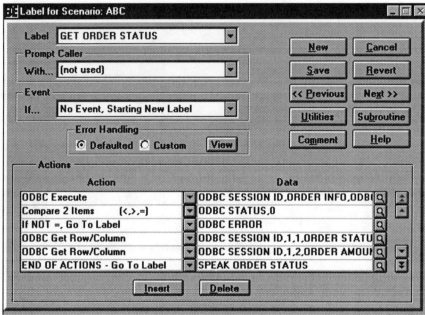

Figure 22-12

VI|
Computer Telephone Integration (CTI)

23|
CTI Overview

24|
TAPI

25|
TSAPI

26|
Versit

27|
CT-Connect

23|
CTI Overview

Computer Telephone Integration (CTI) Basics

CTI focuses on productivity enhancements for people who work with telephones and computers simultaneously. CTI examples include "screen pop" and coordinated call and data transfer. As the name indicates, computer telephone integration is all about getting computer applications and telephone switches to work together. A physical interconnection between the two is essential, so that computer-based applications can request telephone services and track calls through the switch.

Two kinds of physical connections are possible:

- Individual desktop connections between each user's workstation and that same user's telephone; and,

- A single shared connection between a LAN-based server and the telephone switch.

The individual desktop connection is usually best for small numbers of users and for relatively simple applications, such as screen based dialers. A CTI server makes more sense for medium to large work groups, and for more sophisticated applications such as intelligent call routing.

CTI Application Programming Interfaces (APIs)

An API is simply the vocabulary and grammar with which an application makes requests. An API defines a set of function calls that a programmer uses to perform a particular operation. An API is completely local to a particular application machine and is typically integrated with the application that runs on clients. The application cannot access functions and features that the API does not support.

The two most popular CTI APIs are Microsoft's Telephony API (TAPI) and Novell's Telephony Services API (TSAPI). As currently defined, neither API is functionally complete or supports the full range of computer, network and telephone environments. However, interest is high and progress is swift.

First-Party vs. Third-Party Call Control

In CTI applications, telephones can either be controlled individually at each workstation, or remotely by the switch. In general, the way the device is controlled determines the functionality that the application can provide.

First-Party Call Control

The most common example of a first-party call control device is a modem in a workstation used for auto-dialing in conjunction with contact management applications.

First-party call control acts on behalf of one user and usually requires a telephone type-specific circuit board installed in the workstation. The ISA compatible, peripheral card installs in the agent's PC and allows that PC and its application to interact with the associated telephone connection.

While the control of the telephone is enhanced through the PC application, the voice connection is still processed through a traditional handset or headset. A new board is required for each proprietary type of phone.

Desktop
Call Control Interface

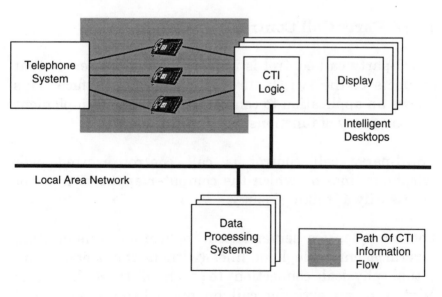

Figure 23-1

The advantage of this type of control is that the switch sees the workstation as an executive or supervisors' telephone. This provides either a super-set of functions or operator console functions to a basic telephone. It provides functions such as call dialing, answer and transfer as well as screen pop via ANI/DNIS, Caller ID or information available on the workstation.

First-party control provides control of a call, but from the user's end point perspective only (not from a detached third party perspective). The disadvantages in this environment are that new boards must be purchased for new telephone types (and added for each new user), the system is limited to the tasks the board can perform, the telephone device must be in the circuit, and most importantly -- call control can only be exerted over calls to which the agent is a "talking" party.

Third-Party Call Control

Third-party call control is implemented via a server that connects to the switch through a digital interface. This allows an application to control the switch and implement a broad array of functions. See Figure 23-2.

Third-party call control is call control exerted over telephone lines on which the computer application is not necessarily a "talking" party.

For example, if a server-based application is monitoring several agents' telephone lines (without the benefit of an actual physical connection to each of those lines), is alerted to an arriving call on one of those lines, and causes that call to be diverted to some other agent's telephone, then the application is exerting third-party call control. Third-part call control usually implies out-of-band signaling, since there is by definition no direct connection between the computer system running the application and the telephone line being controlled.

Server-Based
Call Control Interface

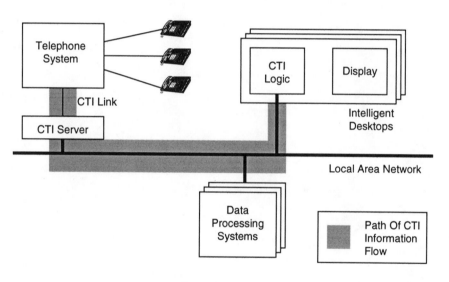

Figure 23-2

The advantage of this approach is that multiple telephones and switching environments can be supported. Enhancements can be made via software. An application can implement a broader array of functionality than with first party control because more features and functionality are typically supported via this digital interface or CTI link. In addition, medium to large groups of user's are more easily, and efficiently supported in this environment.

24|
TAPI

TAPI 1.X

TAPI was designed to do for telephony devices what the Windows Print Manager did for printers. TAPI is an API that was developed by Intel and Microsoft to (at least initially) provide first party call control. The functions that are defined as part of TAPI 1.X specifically address the control of a call from an end point of the call. TAPI 1.X does not contain certain functions that are typically associated with third party call control, such as getting and setting ACD agent states. These extensions are targeted to be added to TAPI 2.0 (addressed later).

TAPI's architecture consists of two components: and API and a Service Provider Interface (SPI). These SPIs are very similar in function to a traditional device driver.

TAPI API

TAPI API function calls are defined by levels of service:

- **Basic** - These functions provide a guaranteed level of service that corresponds to Plain Old Telephone Service (POTS) functions. These functions enable an application to make and receive calls and all must be implemented by the Service Provider.

- **Supplementary** - These function calls provide more advanced levels of service, such as the ability to put a call on hold or transfer. These functions are optional for a Service Provider to implement, however, an application can query to determine which Supplementary functions are supported by the Service Provider.

- **Extended** - These function calls provide access to Service Provider specific functions not directly defined by TAPI. Cross platform capability can be seriously limited by using these functions.

Service Provider Interface (SPI)

The Service Provider Interface (SPI) is the "device driver" component of the TAPI architecture. This interface receives its requests and sends its replies to a dynamic link library (DLL). On the other side of this DLL are the TAPI API functions. The SPI is not provided by Microsoft. The implementation of these functions are provided by the manufacturers of telephony devices (from modems to switches) or by third party developers. This creates the potential for different developers to interpret the specifications differently, or implement functions differently, which can lead to application consistency problems.

One TAPI architectural distinction worth noting is that TAPI contains a "state machine" due to the fact that it was defined from the perspective of a live user operating a physical device. TAPI keeps state information. In the case of a call transfer, a person would flash-hook, dial the number to transfer to, then go on-hook to complete the transfer. TAPI keeps track of these state transitions because that is what a real person would do from their handset to complete a transfer. This may add complexity to an application that needs to manipulate the call from a third party perspective, or for development of a TAPI Service Provider.

TAPI 2.0 Architecture

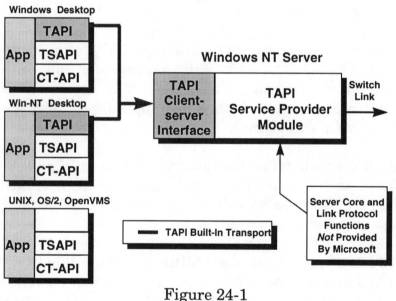

Figure 24-1

TAPI 2.0

Among other service extensions and enhancements, TAPI 2.0 features call center support and third-party call control via a client/server extension. The client side will

365

basically remain the same from an architecture and service level standpoint, but the Service Provider Interface can reside on an NT server. TAPI API requests from an application will be re-directed across a LAN to the NT server. This remote architecture will still require an SPI provided by the switch manufacturer or by third party developers.

Goals of TAPI 2.0

The Win32 implementation of Windows Telephony has the following major goals:

- **Native 32-bit support** - All core TAPI components will be Win32, with full support for non-Intel processors (running Windows NT), symmetrical multiprocessing, multithreaded applications, and pre-emptive multitasking.

- **32-bit application portability** - Existing Win32 full TAPI and assisted TAPI applications which currently run on Windows 95 will run on Windows NT without modification or recompilation.

- **16-bit application portability** - Existing Win16 full TAPI and assisted TAPI applications which currently run on Windows 95 will run on Windows NT without modification or recompilation.

- **32-bit service provider interface** - Telephony Service Providers on Windows NT must be Win32; 16 bit service providers are not guaranteed to run on Windows NT.

- **NDISTAPI compatibility** - The existing support in Windows NT 3.5 for ISDN WAN miniports under Remote Access Service will be preserved. NDIS WAN miniport drivers will be supported under a kernel mode service provider without modification.

- **Registry support** - All telephony parameters will be stored in the registry. Telephony service providers and all stored parameters can be updated across the LAN.

- **Call Center support** - Appropriate enhancements will be made to TAPI and TSPI to support functionality required in a call center environment. This includes the modeling of predictive dialing ports and queues, ACD agent control, station set status control, and centralized event timing.

- **Dialing properties enhancements** - The default dialing rules for countries and calling cards will be editable. Toll lists will also be editable. Users in country code 1 will be able to specify that all local calls should be dialed using 10 digits, and to specify exchanges in other area codes that should be dialed using 10 digits instead of 11.

- **Quality of Service support** - Applications will be able to request, negotiate, and re-negotiate quality of service (performance) parameters with the network, and receive indication of QOS on inbound calls and when QOS is changed by the network. The Quality of Service structures are binary-compatible with those used in the Windows Sockets 2.0 specification.

- **Enhanced Device Sharing** - Applications will be able to restrict their handling of inbound calls on a device to a single address, to support features such as distinctive ringing when used to indicate the expected media mode of inbound calls. Applications making outbound calls will be able to set the device configuration when making a call.

- **Device pools** - Any enhancements to Windows Telephony needed to support device pools (such as server-based pools of modems, fax cards, voice/IVR cards, ISDN lines, digital T1 trunks, etc.) will be provided.

- **User mode components** - The full TAPI system, including top-level service provider DLLs, will run in user mode. It is not a goal to provide direct access to TAPI services to kernel mode components. Service providers can, of course, use any appropriate mechanism to access kernel mode components, should they choose to use such an architecture.

TAPI System Support

TAPI 1.4 is presently supported on Microsoft Windows 95 clients. TAPI 2.0 is supported on Windows NT Server 4.0 and Windows NT Workstation 4.0.

TAPI Pricing

Microsoft currently does not charge for TAPI. It is a standard feature of operating systems in which it is supported. It is unknown whether Microsoft will charge for the remote client/server extension with TAPI 2.0.

TAPI compatible modems are currently available for approximately $250. Switch vendors are currently pricing TAPI 1.X workstation boards at the price of a comparable proprietary telephone ($400-$700). Since TAPI 2.0 SPI's are not yet available, their pricing is unknown. It is anticipated that TAPI 2.0 SPI's will be reasonably priced on new switches. However, upgrading existing switches could entail basic switch software upgrades as well as purchase of the SPI.

Switch Vendors Support for TAPI

The following switch vendors acknowledged participation in the TAPI interoperability test (TAPI Bakeoff) held in Honolulu in May of 1996:

- Comdial

- Ericsson

- Fujitsu

- Harris

- Iwatsu

- Lucent (AT&T)

- Mitel

- NEC

- Nortel (Northern Telecom)

- Rockwell

- SDX

- Toshiba

25|
TSAPI

TSAPI Overview

TSAPI is an API which is based on the Computer
Supported Telephony Applications (CSTA) specification
from the European Computer Manufacturer's Association
(ECMA). TSAPI was developed by Novell and AT&T.

CSTA defines the syntax and protocol with which an
application running on a host computer can obtain and
control telephone switch functions. TSAPI is a subset of
this specification which defines the switch-to-host link
command and status interface.

TSAPI supports third party call control, i.e. the API
functions that it defines are intended to be exercised from
a third party perspective dealing with the "call" as
opposed to the media stream.

Figure 25-1 provides an overview of the Novell Telephony
Services architecture.

TSAPI function calls are defined in the following
categories:

- **Control Services** - These function calls control the
  APIs and the CSTA Control Services. These function
  calls are specific to the Novell Telephony Server

architecture. Applications use these services to manage their interactions with NetWare Telephony Services. The primary function of these services are to open a "stream" or a channel to/from the application and the Telephony Server and to query device and function information from the Telephony Server.

Novell Telephony Services Architecture

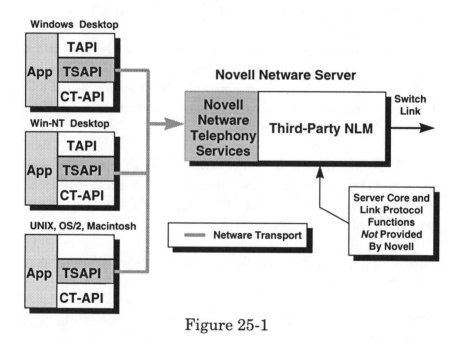

Figure 25-1

- **Switching Function Services** - These function calls are used to control calls and activate switch features. Switching functions are defined by the Basic Call

372

Control Services and the Telephony Supplementary Services. The Basic Call Control Services allow the application to answer, transfer, tear-down calls, and activate and deactivate switch features. The Supplementary Services are used to manipulate Telephony Objects.

- **Status Reporting Services** - These functions are used to provide event status reporting of a CSTA device (such as a handset) to an application. These services can be turned on or off by an application.

- **Snapshot Services** - These functions are used to query the current state of an object or a CSTA call. The information returned from these functions is referred to as a snapshot, since the state of the call or device can change over time.

- **CSTA Computing Functions** - Currently Application Call Routing is the only CSTA Computing Function supported by TSAPI. This is used by an application to supply destinations on a call-by-call basis.

- **Escape and Maintenance Functions** - These functions allow an application to add "private" services which may be specific to a switch or PBX driver and that are not covered as part of the TSAPI definition. Cross platform capability can be seriously limited by using these functions.

- **Network Loadable Modules (NLMs)** - These are equivalent to the TAPI SPI and are basically the device driver for the physical medium connected to the

Novell Telephony Server. These NLMs are not supplied by Novell (except the AT&T NLM) but developed by switch vendors or third party developers. This creates the potential for different developers to interpret the specifications differently, or implement functions differently, which can lead to application consistency problems.

In comparison to TAPI, the TSAPI implementation does not specifically keep state machine information like TAPI does. This API abstracts the CSTA call model of the switch, enabling an application to control calls and devices of the switch without explicitly keeping track of the switch's states.

TSAPI System Support

TSAPI presently supports a Novell NetWare server with TCP/IP or IPX/SPX environment. Supported clients include Microsoft Windows, Win95 and NT as well as UnixWare, OS/2 and Macintosh.

TSAPI Pricing

Netware Telephony Services version 2.21 is currently available and is integrated with NetWare 4.1. Licenses are available for 5, 10, 25, 50, 100 and 250 users at prices ranging from $1,295 for 5 users to $26,995 for 250 users.

Switch vendors are not consistent in their pricing of the TSAPI option. Fujitsu charges approximately $10,000 for their TSAPI option. Upgrading existing switches could entail basic switch software upgrades as well as purchase of the TSAPI option.

Switch Vendors Support for TSAPI

37 PBX vendors are actively developing or have already developed drivers for NetWare Telephony Services including:

- Alactel

- BBS Telecom

- Comdial

- Cortelco

- Ericsson

- Fujitsu

- Hitachi

- Intecom

- Intertel

- Iwatsu

- Lucent (AT&T)

- Mitel

- NEC

- Nortel (Northern Telecom)

- OKI

- Panasonic

- Philips

- Rockwell

- SDX

- Siemens ROLM

- SRX

- Tadiran

- Telrad

- Toshiba

26|
Versit

About Versit

Versit is a global initiative of Lucent Technologies, IBM, and Siemens which seeks to eliminate barriers to communication and collaboration by developing and promoting open, cross-platform specifications.

The Versit CTI initiative is built on the following foundation technologies:

- CSTA Standards by ECMA (European Computer Manufactures Association)

- CallPath Normalization by IBM

- TSAPI by Novell/AT&T

Versit CTI Encyclopedia

Versit has published the *Versit CTI Encyclopedia - Release 1.0* defining terminology, configurations, CSTA feature set, call flows, protocols and Versit TSAPI. Key features of the document include:

- Call Control Services

- Device Control Services

- Direct-Connect Configurations

- Client-Server Configurations

- First Party Call Control

- Third Party Call Control

- Media Services

- Routing Services

- Monitoring and Status Reporting

The standards based specifications define a broad, consistent call model, an open API, and transport independent protocols. By defining operations independent of connectivity and operating systems or hardware platforms, the open framework enables true cross-platform interoperability. It allows the development of functionally-rich applications that work consistently in either direct-connect or client-server configurations.

Direct-Connect Configuration

The Versit Direct-Connect configuration involves a telephone set and a client, which may be a PC or a PDA (Personal Digital Assistant). The connection may be a cable, infrared, add-on card, or any other physical transport. See Figure 26-1.

Versit Direct-Connect Configuration

Figure 26-1

Client-Server Configuration

The Versit Client-Server configuration involves a telephony server which interacts with a switch on behalf of one or more clients.

Versit Client-Server Configuration

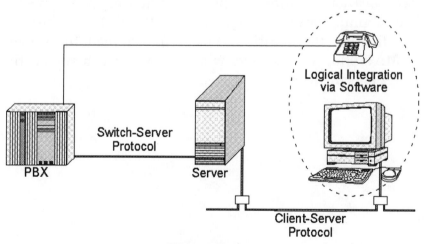

Figure 26-2

Although there is no physical connection between the telephone set and the client workstation, Versit defined protocols communicate between the client and the server, and between the server and the switch. See Figure 26-2.

Versit TSAPI

Versit TSAPI is a procedural API specification based on the concepts, terminology, configurations and telephony feature set defined in the Versit CTI Encyclopedia. It is platform independent - which means that it attempts to remove the Novell Netware and AT&T Definity biases from the original TSAPI while allowing TSAPI developers to adapt their existing applications to the new API.

Versit and ECMA

Versit recently gave its CTI Encyclopedia to ECMA for incorporation into the Computer Supported Telecommunications Application (CSTA) Phase III specification. In addition, Versit has agreed to support ECMA's efforts to achieve wide-spread agreement on the CSTA specification

27|
CT-Connect

The API Dilemma

In addition to defining their respective API's, both Novell and Microsoft have also defined system architectures for CTI servers. Not surprisingly, each vendor has established a server architecture that emphasizes their own products, reducing your flexibility once you install a server of their recommended architecture.

For example, the Novell CTI server architecture intentionally only supports the TSAPI interface and requires the presence of Netware Directory Services; the Microsoft architecture supports only the TAPI interface and requires all applications to run under the Microsoft Windows family of operating systems. A CTI software vendor who conforms strictly to one of these designs cannot support multiple APIs or a truly open computing and networking environment.

Most switch vendors have followed this strategy, offering software modules which fit into the Novell or Microsoft architecture - but not a single module that does both. With this either/or approach, you are faced with making a commitment to one of the two architectures.

If you install the Novell-style module on a Netware server, you may not have support for TAPI-based applications. If you install the Microsoft-style module on a Windows NT server, you may not have support for TSAPI-based applications or non-Windows client workstations.

What Is CT-Connect?

CT-Connect is an open, standards-based computer-telephone link server from Dialogic Corporation. One of the unique features of CT-Connect is its support for multiple telephony programming interfaces, including the Intel/Microsoft Telephony Application Programming Interface (TAPI), the Novell/AT&T Telephony Services Application Programming Interface (TSAPI), and the Microsoft Dynamic Data Exchange (DDE) interface. In particular, CT-Connect has a unique ability to support desktop TAPI applications in server-based work-group environments where a common server approach is more cost-effective than individual desktop telephone connections.

Getting the Architectural Labels Right

The publicity surrounding CTI products does not always clearly label the product components according to accepted system architecture. For example, product literature and press releases do not always draw a clear line between design elements such as:

- Programming interface specifications, such as TAPI or TSAPI;

- Communication mechanisms between application machines and CTI server machines, such as TCP/IP, IPX/SPX, Winsock, or RPC;

- Commercial CTI server implementations, such as CT-Connect or Novell's Netware Telephony Services; and,

- Communications protocols used between the CTI server and a specific switch, such as the ECMA CSTA protocol, AT&T's ASAI protocol, or Northern Telecom's Meridian Link protocol.

All too often, a CTI vendor will blur these architectural elements -- for example, by claiming that a client-server connection is "CSTA compliant" when in fact the CSTA standard has nothing whatsoever to do with client-server communications. (The CSTA protocol specification addresses only the information flow between the switch and the computer system it is physically wired to.) Or a vendor might combine certain of these elements into a single un-modifiable package, by restricting the implementations of a certain API to customers using a specific network or server product.

In Dialogic's view, a customer should be free to choose independently among these categories -- for example, to choose a commercial CTI server implementation without thereby restricting the range of LAN products that can be used, or to choose a desktop application package (and thus implicitly a CTI API) without thereby being forced to use a particular CTI server product. In the design of CT-Connect, Dialogic has tried very hard to maintain this openness and freedom of choice. This is why CT-Connect is described as an "open" CTI server. See Figure 27-1.

CT-Connect Architecture

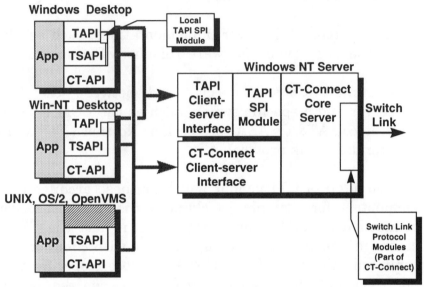

Figure 27-1

CT-Connect and API Support

As Dialogic designed CT-Connect, they anticipated from the very first that they would need to support several APIs. They knew they would need a full-featured API to support sophisticated applications. So they implemented a CSTA-based CT-Connect API that made the greatest possible range of switch link capabilities available to the application program. Such an API is more complex than most applications need, so they knew we would also have to support other popular APIs.

Of course, it is not essential (or likely) that the entire world agree on a single telephony API. Programming interfaces are like most other areas of technology; one size hardly ever fits all. And telephony APIs are like programming languages. We have more than we really want, but there are valid, unique advantages to many of them. And as long as they can interoperate on a common computing platform, who really cares? So as they designed CT-Connect, they have always kept open the possibility of supporting multiple APIs.

As long as each new API can be defined within the existing CSTA-based telephony resource model, adding a new API to CT-Connect is not difficult. So if other APIs become commercially popular, and if they can be implemented on an open environment (for example, without requiring specific commercial LAN or CTI server products), Dialogic can support them under CT-Connect.

How CT-Connect Supports Multiple APIs

Remember that an API is completely local to a particular application machine. An API is not a network concept, and the API does not appear at all on the CTI server. It is simply the vocabulary with which the application makes CTI requests. And remember that the application cannot see through the API to figure out what lies behind it; whether the resource manager behind the API is simple or complex, the application sees the same thing: just the API itself.

Each application system that needs to connect to the CT-Connect server uses a client module supplied with CT-Connect. Client modules are available to support applications running under a range of operating systems:

- Microsoft Windows

- Microsoft Windows NT

- Several varieties of Unix

- Digital OpenVMS

- OS/2

The client modules implement one or more APIs within the target operating system and handle the client-server message traffic to the CT-Connect server. Of course, many client systems and their corresponding applications can be simultaneously connected to a single CT-Connect server.

For example, CT-Connect now has client modules for the 16-bit Microsoft Windows family which provide:

- CT-Connect API

- Microsoft TAPI interface

- Novell TSAPI interface

- Microsoft DDE interface

This gives a customer the choice of virtually any of the available CTI applications, and several applications using different telephony APIs can operate on the same desktop at the same time.

Network Protocol Independence

Any client-server CTI architecture needs a mechanism for getting CTI requests from application systems to the server, and for relaying the results from the server back to the application system. While other CTI products on the market restrict a customer to particular network protocols, CT-Connect allows complete network flexibility by supporting two standard network services: the OSF DCE (Distributed Computing Environment) specification and Winsock (the Windows socket library specification).

For environments other than Windows 3.1, a customer can use any OSF DCE compliant network operating system to link the application machines and the CT-Connect server. For Windows 3.1, whose limited multi-tasking does not support OSF DCE networks well, CT-Connect operates with any Winsock-compliant network.

CTI as an Enterprise-Wide Resource

In a typical configuration, a customer will often have a mixture of PC-equipped desktops, dumb-terminal desktops, time-sharing or mainframe hosts, true client-server database servers, voice response applications, LANs, WANs, and so forth. More advanced environments may require multi-site telephony services to allow call-data transfers across sites. If so, a customer will need telephony services that operate across both workgroup and corporate-backbone data network technologies.

Supporting this mixture of data processing and networking resources requires a truly flexible CTI architecture, one which does not introduce its own

387

environmental restrictions. This is the only way for a customer to make telephony services available uniformly across an enterprise and to preserve their investment in computer-telephone technology.

Dialogic believes that CTI capabilities should be designed and implemented as an enterprise-wide capability because that is how they will inevitably be used. It is difficult to predict today where computer telephone capabilities may be needed tomorrow.

CT-Connect was designed to be a universal enterprise-wide CTI server, integrating easily with almost any software, operating system, computing hardware or network environments. CT-Connect's support for TAPI and TSAPI is part of that design, ensuring that popular desktop CTI applications will be able to share enterprise computer-telephone resources with other applications (which may employ different telephony APIs), and allowing a customer to utilize the same applications at both desktop-connect sites as well as server-supported sites.

Switch Links Supported by CT-Connect

The standard version of CT-Connect supports switches with CSTA-compliant CTI links, as well as some additional switches whose CTI links are similar to the CSTA call model.

The following switch links are supported by the standard version:

- Alcatel 4400 (CSTA Link)

- AT&T Definity G3 (CallVisor ASAI)

- BBS Telecom 308/416 (CSTA Link)

- Ericsson MD110 (ApplicationLink)

- Northern Telecom Meridian 1 (Meridian Link)

- Siemens Rolm 9751 CBX (CallBridge/CSTA)

Special versions of CT-Connect support the following proprietary CTI links:

- Aspect Call Center (ApplicationBridge)

- Rockwell Galaxy (Transaction Link)

- Siemens HICOM 300 (CallBridge/ACL)

Configurations and Pricing

The standard version of CT-Connect is available in three commercial configurations:

- Full configuration - supporting all types of client modules and an unlimited number of client systems

- Desktop configuration - supporting an unlimited number of client systems but restricted to desktop applications

- Desktop Lite configuration - supporting up to 36 client systems running desktop applications

Pricing ranges from $2,850 to $14,250 for quantity one and excludes the cost of any necessary switch upgrades or CTI link options.

VII|
Conclusion

28|
The Book in Review

28|
The Book in Review

Summary

Whether this book was your first formal introduction to Computer Telephony or an adjunct to your ever expanding knowledge and understanding of the technology, *Understanding Computer Telephony* will become a well referenced addition to your library. As with any emerging technology, changes are inevitable and exciting. This book explains the fundamental technologies as well as many of the advanced capabilities being incorporated into CT solutions every day.

In Section I, you were exposed to the reasons and justifications why you as a CT developer and provider will undoubtedly be successful with this emerging technology. The most successful VARs and developers to date have focused on vertical markets and/or specific applications with their CT solutions. If you already have an existing customer base, it would definitely be the best place to begin your marketing.

As you focus on your vertical market(s) and/or applications, an important realization is that similar CT applications actually can be marketed to many varying industries. For example, the application that was used in this book is for obtaining an order status with options for hearing daily specials. There are countless organizations providing products and services that can benefit from this type of application. Whether your customers are selling pizzas, houses, or jet engine parts, they probably have full time employees responding to incoming calls for an order status, account balance, and more. The benefits to you as a CT provider are minimized development time and unlimited opportunities.

This same premise holds true for numerous other applications. For example, many organizations attempt to gather critical marketing information to support their products or services. A CT application for a hotel chain designed to capture guest feed back regarding their recent stay could also be used by automobile dealers, computer manufacturers, or the local tourist bureau. The idea is to let your imagination explore all the opportunities for your completed application development.

As you progressed through Section II, the various telephone network connections were detailed. Don't let this confuse you. The objective is to make you aware of the various interfaces and their capabilities. While your systems will primarily be analog based, digital network interfaces will play a major role when higher line density opportunities come along.

Section III was a must read because CT systems have to receive input from the callers. Once again, most of your opportunities will utilize touchtone/DTMF detection. When you are working in the analog interface environment, caller ID may add a distinct advantage to your application and help to win the business. Also, a working knowledge of Automatic Speech Recognition (ASR) will allow you to converse at a level that your competition probably can not. Although you may not implement ASR in the near future, chances are, this technology will be requested, and you will eventually incorporate ASR into a CT solution.

Section IV detailed the various methodologies for prompting the system callers and delivering requested information through voice or fax, and to the hearing impaired. The number one (but oftentimes overlooked) issue with any system you implement is the quality of the information delivery. Many CT systems don't place a priority on the quality of the voice recording. If you think about it, what the caller hears or receives by fax accounts for 90% of the overall impression of the system. Don't minimize the importance of high quality voice playback. You can save some money in the short run but it will cost you business in the long run. The main objective in any system is to meet your customers' specific needs.

Section IV also reviewed call transfer and out-dialing. Again, the ability to transfer to a live operator is critical to any system that is used for customer service type applications. Make sure that your system design allows for this capability, and convince your customers that the success of their systems depends on providing this option.

Section V dealt with the not so clear aspect of accessing data. Don't hesitate to rely on out-sourced expertise if you don't have a clear understanding of this area. Your customers may even be able to provide the knowledge necessary to integrate your CT system with their hosts and data sources. A word of caution, users are much more patient with a data system than with a voice system. A few seconds waiting for an audible response regarding inventory over the phone seems like an eternity compared to staring at a terminal for a displayed response.

Finally, Section VI provided an overview of Computer Telephone Integration (CTI). This introduced the capability of screen pop and coordinated call and data transfer. Current activity in this area is focused on the "API war" with all the attendant speculation and hedging. Although the future is not clear in this area, interest is high and progress is swift.

Implementing the CT System - Buy or Build?

Understanding Computer Telephony focuses on CT technologies, and shows numerous examples of how the technologies can be implemented with the EASE application development tool. As you may know, applications such as voice mail, fax-on-demand, and many others are available on the open market for resale. These applications, although feature rich in most cases, are still "closed" applications that limit your ability to enhance or add functionality above and beyond the specific application. Vendors like Expert Systems offer application templates (called EASEy FrameWorks) that are coded applications utilizing an application development tool. These templates can be used to "jump-start" your application development.

Functional for demonstration purposes, templates save a tremendous amount of development time without limiting you to added functionality.

In reviewing your key business objectives for implementing a CT system, you should keep five basic goals in mind:

⇒ Enhance the business operation
⇒ Reduce costs
⇒ Generate new revenue opportunities
⇒ Improve customer service
⇒ Gain a competitive edge.

If you are able to meet these primary goals, your customers will be satisfied, and your CT venture will be successful.

Computer Telephony Software Market

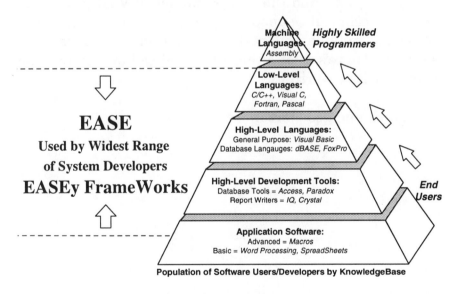

Figure 28-1

As you've learned, deploying CT system solutions is based on strategic and economic reasons and is not a technology issue. Choosing the right tool is one of the first major decisions you have to make. To help you make that

397

decision, figure 28-1 illustrates the potential skill sets and user base for various programming tools. It also shows where a development environment like EASE and application templates such as EASEy FrameWorks meet the outlined requirements.

In conclusion, developing CT applications can be accomplished with an ever growing array of tools, library extensions, and of course in C++. Many developers choose to develop with a programming language such as Visual Basic, while others decide on turnkey applications. Before you make your decision, it's important to look at the company behind the tool. Established VAR programs, skilled technical support, and extensive training programs provide the best combination to ensure your success. There's a lot of hard work ahead of you, but the results will be worth it.

Good luck and good fortune!

Glossary|

23B+D

An easy way of saying the ISDN Primary Rate Interface circuit. 23B+D has 23 64 Kbps (kilobits per second) paths for carrying voice, data, video or other information and one 64 Kbps channel for carrying out-of-band signaling information. ISDN PRI bears a remarkable similarity to today's T-1 line, except that T-1 can carry 24 voice channels. In ISDN, 23B+D the one D channel is out-of-band signaling. In T-1, signaling is handled in-band using robbed bit signaling. Increasingly, 23B+D is the preferred way of getting T-1 service since the out of band signaling is richer (delivers more information -- like ANI and DNIS) and is more reliable than the in-band signaling on the older T-1. See ISDN, PRI, ROBBED BIT SIGNALING and T-1.

2500 SET

The normal single-line touchtone desk telephone. It has replaced the rotary dial 500 set in most, but definitely not all, areas of the US and Canada.

3270

IBM class of terminals (or printers) used in SNA mainframe environments.

5250
IBM class of terminals (or printers) for midrange (System 3x and AS/400) environments.

A & B BIT SIGNALING
Procedure used in most T-1 transmission links where one bit, robbed from each of the 24 subchannels in every sixth frame, is used for carrying dialing and controlling information. A type of in-band signaling used in T-1 transmission. A and B signaling reduces the available user bandwidth from 1.544 Mbps to 1.536 Mbps.

ACD
See AUTOMATIC CALL DISTRIBUTOR

ADAPTIVE DELTA PULSE CODE MODULATION
A way of encoding analog voice signals into digital signals by adaptively predicting future encodings by looking at the immediate past. The adaptive part reduces the number of bits per second that another rival and more common method called PCM (Pulse Code Modulation) requires to encode voice. Adaptive PCM is not common because, even though it reduces the number of bits required to encode voice, the electronics to do it are expensive. See PULSE CODE MODULATION.

ADPCM
See ADAPTIVE DELTA PULSE CODE MODULATION

AID

Attention IDentification for initiating communication between a terminal and an IBM mainframe and midrange host system.

A-LAW

The PCM coding and companding standard used in Europe. The method of encoding sampled audio waveforms used in the 2.048Mbit 30 channel PCM primary system, widely used outside North America. See also COMPANDING.

ALIASING

Distortion in a signal. In video, it shows up in different ways depending on the type of aliasing in question. When the sampling rate interferes with the frequency of program material the aliasing takes the form of artifact frequencies that are known as sidebands. Spectral aliasing is caused by interference between two frequencies such the luminance and chrominance signals. It appears as herringbone patterns, wavy lines where straight lines should be and lack of color fidelity. Temporal aliasing is caused when information is lost between line or field scans. It appears when a video camera is focused on a CRT and the lack of scanning synchronization produces a very annoying flickering on the screen of the receiving device.

AMERICAN NATIONAL STANDARDS INSTITUTE

American National Standards Institute. A standards-setting, non-government organization, which develops and publishes standards for "voluntary" use in the United

401

States. The British have their own equivalent of ANSI. It's called the BSI, British Standards Institute. Standards set by national organizations are accepted by vendors in that country. ANSI is located at 1430 Broadway, New York NY 10018 212-642-4900. They put out a biweekly newsletter called "ANSI Standards in Action."

In a press release, ANSI described itself as "a private non-profit membership organization that coordinates the U.S. voluntary standards system, bringing together interests from the private and public sectors to develop voluntary standards for a wide array of U.S. industries. ANSI is the official U.S. member body to the world's leading standards bodies -- the International Organization for Standardization and the International Electronic Commission via the U.S. National Committee. The Institute's membership includes approximately 1,300 national and international companies, 30 government agencies, 20 institutions and 250 professional, technical, trade, labor and consumer organizations."

AMPHENOL CONNECTOR
Amphenol is a manufacturer of electrical and electronic connectors. They make many different models, many of which are compatible with products made by other companies. Their most famous connector is probably the 25-pair connector used on 1A2 key telephones and for connecting cables to many electronic key systems and PBXs. The telephone companies call the 25-pair Amphenol connector used as a demarcation point the RJ-21X. The RJ-21X connector is made by other companies including 3M, AMP and TRW. People in the phone business often call non-amphenol 25-pair connectors, amphenol connectors.

ANALOG

Comes from the word "analogous," which means "similar to." In telephone transmission, the signal being transmitted -- voice, video, or image – is " analogous" to the original signal. In other words, if you speak into a microphone and see your voice on an oscilloscope and you take the same voice as it is transmitted on the phone line and ran that signal into the oscilloscope, the two signals would look essentially the same. The only difference is that the electrically transmitted signal (the one over the phone line) is at a higher frequency. In correct English usage, "analog" is meaningless as a word by itself. But in telecommunications, analog means telephone transmission and/or switching which is not digital. See ANALOG TRANSMISSION.

ANALOG TRANSMISSION

A way of sending signals -- voice, video, data -- in which the transmitted signal is analogous to the original signal. In other words, if you spoke into a microphone and saw your voice on an oscilloscope and you took the same voice as it was transmitted on the phone line and threw that signal onto the oscilloscope, the two signals would look essentially the same. The only difference would be that the electrically transmitted signal would be at a higher frequency.

ANI

See AUTOMATIC NUMBER IDENTIFICATION.

ANSI

See AMERICAN NATIONAL STANDARDS INSTITUTE

ANTI-ALIASING

A computer imaging term. A blending effect that smoothes sharp contrasts between two regions of different colors. Properly done, this eliminates the jagged edges of text or colored objects. Used in voice processing, anti-aliasing usually refers to the process of removing spurious frequencies from waveforms produced by converting digital signals back to analog.

ASCII

Pronounced: as'-kee. American Standard Code for Information Interchange. It's the most popular coding method used by small computers for converting letters, numbers, punctuation and control codes into digital form. (Computers can only understands zeros or ones.) Once defined, ASCII characters can be recognized and understood by other computers and by communications devices. ASCII represents characters, numbers, punctuation marks or signals in seven on-off bits. A capital "C", for example, is 1000011, while a "3" is 0110011. As a seven-bit code, and since each bit can only be a "one" or a "zero,"

ASCII can represent 128 "things," i.e. 2 x 2 x 2 x 2 x 2 x 2 x 2 which equals 128. ASCII is the code virtually every personal computer in the world encodes "things," including IBM, Apple and Radio Shack/Tandy. This compatible encoding (it was developed by ANSI -- the American National Standards Institute) allows virtually all personal computers to talk to each other, if they use a compatible modem, or null modem cable and transmit and receive at the same speed. There are variations of ASCII. (Nothing is totally standard anymore.) The most important variation – one originally from IBM -- is called

Extended ASCII. It codes characters into eight bits (or one byte) and uses those ASCII characters above 127 to represent foreign language letters, and other useful symbols, such as those to draw boxes. But at 127 and below, extended 8-bit ASCII is identical to standard 7-bit ASCII. The CCITT (now called the ITU-T) calls ASCII International Telegraph Alphabet 5.

The other major method of encoding is IBM's EBCDIC (pronounced ebb'-si-dick). It's largely used on IBM and IBM-compatible mainframe computers (but not their PCs, which use ASCII and extended ASCII.) EBCDIC is an eight-bit encoding scheme, thus allowing up to 256 "things" to be encoded, i.e. 2 x 2 x 2 x 2 x 2 x 2 x 2 x 2 = 256. EBCDIC codes letters, characters and punctuation marks in a totally different way than ASCII. For ASCII files to be read by an IBM mainframe (one that reads EBCDIC), those ASCII files must be translated into EBCDIC by one of the many translation programs available.

ASR
See AUTOMATIC SPEECH RECOGNITION

ASYNCHRONOUS TRANSMISSION
Literally, not synchronous. A method of data transmission which allows characters to be sent at irregular intervals by preceding each character with a start bit, and following it with a stop bit. It is the method most small computers (especially PCs) use to communicate with each other and with mainframes today. In every form of data transmission, every letter, number or punctuation mark is transmitted digitally as "ons" or "offs." These characters

are also represented as "zeros" and "ones" (See ASCII). The problem in data transmission is to define when the letter, the number or the punctuation mark begins. Without knowing when it begins, the receiving computer or terminal won't be able to figure out what the transmission means.

One way to do this is by using some form of clocking signal. At a precise time, the transmission starts, etc. This is called SYNCHRONOUS TRANSMISSION. In ASYNCHRONOUS transmission there's no clocking signal. The receiving terminal or computer knows what's what because each letter, number or punctuation mark begins with a start bit and ends with a stop bit. Transmission of data is called SYNCHRONOUS if the exact sending or receiving of each bit is determined before it is transmitted or received. It is called ASYNCHRONOUS if the timing of the transmission is not determined by the timing of a previous character.

ASYNCHRONOUS is used in lower speed transmission and by less expensive computer transmission systems. Large systems and computer networks typically use more sophisticated methods of transmission, such as SYNCHRONOUS or BISYNCHRONOUS, because of the large overhead penalty of 20% in ASYNCHRONOUS transmission. This is caused by adding one start bit and one stop bit to an eight bit word -- thus 2 bits out of ten.

The second problem with large transfers is error checking. The user sitting in front of his own screen checks his ASYNCHRONOUS transmission by looking at the screen and re-typing his mistakes. This is impractical for transferring long files at high speed if there is not a person in attendance.

406

In SYNCHRONOUS transmission start and stop bits are not used. According to the book Understanding Data Communications, characters are sent in groups called blocks with special synchronization characters placed at the beginning of the block and within it to ensure that enough 0 to 1 or 1 to 0 transitions occur for the receiver clock to remain accurate. Error checking is done automatically on the entire block. If any errors occur, then the entire block is retransmitted. This technique also carries an overhead penalty (nothing is free), but the overhead is far less than 20% for blocks or more than a few dozen characters.

ATTENUATION
The decrease in power of a signal, light beam, or lightwave, either absolutely or as a fraction of a reference value. The decrease usually occurs as a result of absorption, reflection, diffusion, scattering, deflection or dispersion from an original level and usually not as a result of geometric spreading, i.e., the inverse square of the distance effect. Optical fibers have been classified as high-loss (over 100 dB/km), medium-loss (20 to 100 dB/km), and low-loss (less than 20 dB/km). In other words, attenuation is the loss of volume during transmission. The received signal is lower in volume than the transmitted signal due to losses in the transmission medium (such as that caused by resistance in the cable). Attenuation is measured in decibels. It is the opposite of Gain. Some electrical components are listed as "with attenuation" which means they will compensate for irregular electrical supply (e.g. surges). See GAIN.

AUTO FAX TONE

Also called CNG, or Calling Tone. This tone is the sound produced by virtually all Group 3 fax machines when they dial another fax machine. CNG is a medium pitch tone (1100 Hz) that lasts 1/2 second and repeats every 3 1/2 seconds. A FAX machine will produce CNG for about 45 seconds after its dials. See also CNG.

AUTOMATED ATTENDANT

A device which is connected to a PBX. When a call comes in, this device answers it and says something like, "Thanks for calling the ABC Company. If you know the extension number you'd like, enter that extension now and you'll be transferred. If you don't know your extension, press "0" (zero) and the operator will come on. Or, wait a few seconds and the operator will come on anyway." Sometimes the automated attendant might give you other options, such as, "dial 3" for a directory of last names and dial 4 for a directory of first names. Automated attendants are also connected also to voice mail systems ("I'm not here. Leave a message for me."). Some people react well to automated attendants. Others don't. A good rule to remember is before you spring an automated attendant on your people, customers, subscribers, etc., let them know. Train them a little. Ease them into it. They'll probably react more favorably than if it comes as a complete surprise. The first impression is rarely forgotten, so try to make it a good experience for the caller.

AUTOMATIC CALL DISTRIBUTOR

ACD. A specialized phone system designed originally for handling many incoming calls, now increasingly used by companies also making outgoing calls. You receive and make lots of phone calls typically to customers. You need an ACD. Once used only by airlines, rent-a-car companies, mail order companies, hotels, etc., it is now used by any company that has many incoming calls (e.g. order taking, dispatching of service technicians, taxis, railroads, help desks answering technical questions, etc.). There are very few large companies today that don't have at least one ACD.

An ACD performs four functions. 1. It will recognize and answer an incoming call. 2. It will look in its database for instructions on what to do with that call. 3. Based on these instructions, it will send the call to a recording that "somebody will be with you soon, please don't hang up!" or to a voice response unit (VRU). 4. It will send the call to an agent as soon as that operator has completed his/her previous call, and/or the caller has heard the canned message.

The term Automatic Call Distributor comes from distributing the incoming calls in some logical pattern to a group of operators. That pattern might be Uniform (to distribute the work uniformly) or it may be Top-down (the same agents in the same order get the calls and are kept busy. The ones on the top are kept busier than the ones on the bottom). Or it may be Specialty Routing, where the calls are routed to answerers who are most likely to be able to help the caller the most. Distributing calls logically is the function most people associate with an ACD, though it's not the most important.

The management information which the ACD produces is much more valuable. This information is of three sorts: 1. The arrival of incoming calls (when, how many, which lines, from where, etc.) 2. How many callers were put on hold, asked to wait and didn't. This is called information on abandoned calls. This information is very important for staffing, buying lines from the phone company, figuring what level of service to provide to the customer and what different levels of service (how long for people to answer the phone) might cost. And 3. Information on the origination of the call. That information will typically include ANI (Automatic Number Identification -- picking up the calling number and DNIS (Direct Number Identification Service) picking up the called number. Knowing the ANI allows the ACD and its associated computer to look up the caller's record and thus offer the caller much faster service. Knowing the DNIS may allow the ACD to route the caller to particular agent or keep track of the success of various advertising campaigns. Ad agencies will routinely run the same ad in different towns using different 800 phone numbers. Picking up which number was called identifies which TV station the ad ran on.

AUTOMATIC NUMBER IDENTIFICATION
A phone call arrives at your home or office. Somewhere in that phone call is a series of digits which tell you the phone number of the phone calling you. These digits may arrive in analog or digital form. They may arrive as touchtone digits inside the phone call or in a digital form on the same circuit or on a separate circuit. You will need some equipment to decipher the digits AND to do "something" with them. That "something" might be throwing them into a database and bringing your customer's record up on a screen in front of your telephone agent as he answers the call. "Good morning, Mr. Smith." Some large users say they could save as much as 30

seconds on the average IN-WATS call if they knew the phone number of the person calling them. They would avoid asking regular customers for routine identification information since it would all be there in the database. ANI is touted as one of ISDN's most compelling advantages -- but it is really an advantage of Signaling System 7 (and therefore distinct from ISDN) and you don't need ISDN to get ANI. In the US, there are various types of "ANI." There's the ANI you get from a long distance phone company, which may arrive over the D channel of an ISDN PRI circuit or on a dedicated single line before the first ring. In contrast, the signaling for Caller ID, as delivered by a local phone company, is delivered between the first and second rings. In Canada, caller ID for both local and long distance is delivered in the same technical way -- between the first and second rings. In the US, there are no accepted standards, as yet. In November of 1995, local phone companies in the US are scheduled to deliver both local and long distance ANI exactly as they do today in Canada -- between the first and second rings. Thus normal dial-up users, who subscribe to caller ID (and usually pay a few extra dollars a month for the privilege) will be able to figure who's calling before, or as they pick up the incoming call. At one stage, ANI was not available in many states. But those restrictions are disappearing. There are some people who believe ANI is long distance and delivered by long distance phone companies; and Caller ID is local and delivered by local phone companies. And these same people believe the technologies of delivery are different. In reality, ANI and Caller ID are rapidly becoming synonymous. See CALLER ID, DNIS and ISDN.

AUTOMATIC SPEECH RECOGNITION

Voice recognition is the ability of a machine to recognize your particular voice. This contrasts with speech recognition, which is different. It is the ability of a machine to understand human speech -- yours and everyone else's. Voice recognition needs training. Speaker independent recognition does not require training.

B CHANNEL

A "bearer" channel is a fundamental component of ISDN interfaces. It carries 64,000 bits per seconds in both directions, is circuit switched and is able to carry either voice or data. Whether it does or not depends on how your local telephone company has tariffed its ISDN service. See ISDN.

BASIC RATE INTERFACE

BRI. There are two "interfaces" in ISDN: BRI and PRI. In BRI, you get two bearer B-channels at 64 kilobits per second and a data D-channel at 16 kilobits per second. The bearer B-channels are designed for PCM voice, video conferencing, group 4 facsimile machines, or whatever you can squeeze into 64,000 bits per second full duplex. The data D-channel is for bringing in information about incoming calls and taking out information about outgoing calls. It is also for access to slow-speed data networks, like videotex, packet switched networks, etc. One BRI standard is the "U" interface, which uses two wires. Another BRI standard is the "T" interface which uses four wires. See ISDN for a much fuller explanation.

BAUD RATE

A measure of transmission speed over an analog phone line -- i.e. a common POTS line. (POTS stands for Plain Old Telephone Service). Imagine that you want to send digital information (say from your computer) over a POTS phone line. You buy a modem. A modem is a device for converting digital on-off signals, which your computer speaks, to the analog, sine-wave signals your phone line "speaks." For your modem to put data on your phone line means it must send out an analog sine wave (called the carrier signal) and change that carrier signal in concert with the data it's sending. Baud rate measures the number of number of changes per second in that analog sine wave signal. According to Bell Labs, the most changes you can get out of a 3 KHz (3000 cycles per second) voice channel (which is what all voice channels are) is theoretically twice the bandwidth, or 6,000 baud.

Baud rate is often confused with bits per second, which is a transfer rate measuring exactly how many bits of computer data per second can be sent over a telephone line. You can get more data per second -- i.e. more bits per second -- on a voice channel than you can change the signal. You do this through the magic of coding techniques, such as phase shift keying. Advanced coding techniques mean that more than one bit can be placed on a baud, so to speak. To take a common example, a 9,600 bit per second modem is, in reality, a 2,400 baud modem with advanced coding such that four bits are impressed on each baud. The continuing development of newer and newer modems point to increasingly advanced coding techniques, bringing higher and higher bit per second speeds. My latest modem, for example, is 28,800 bits per second.

BAUDOT CODE

The code set used in Telex Transmission, named for French telegrapher Emile Baudot (1845-1903) who invented it. Also known by the CCITT approved name, International Telegraph Alphabet 2. The Baudot code has only five bits, meaning that only 32 separate and distinct characters are possible from this code, i.e. 2 x 2 x 2 x 2 x 2 equals 32. By having one character called Letters (usually marked LTRS on the keyboard) which means "all the characters that follow are alphabetic characters," and having one other key called Figures (marked FIGS), meaning "all characters that follow are numerals or punctuation characters," the Baudot character set can represent 52 (26 x 2) printing characters. The characters "space," "carriage return," "line feed" and "blank" mean the same in either FIGS or LTRS. TDD devices (Telecommunications Devices for the Deaf) use the Baudot method of communications to communicate with distant TDD devices over phone lines. See also ASCII and EBCDIC, which are other ways of encoding characters into the ones or zeros needed by computers.

BELLCORE

Bell Communications Research. Formed at Divestiture to provide certain centralized services to the seven Regional Bell Holding Companies (RBOCs) and their operating company subsidiaries. Also serves as a coordinating point for national security and emergency preparedness, and communications matters of the federal government. Bellcore does not work on customer premise equipment (e.g. the telephone set) or other areas of potential competition between its owners -- the seven Regional Bell Operating Companies. It is a key player in the design of AIN -- the Advanced Intelligent Network. You can acquire

Bellcore documents from Bellcore -- Document Registrar, 445 South Street, Room 2J-125, P.O. Box 1910, Morristown, NJ 07962-1910. Fax 201-829-5982.

BINARY
Where only two values or states are possible for a particular condition, such as "ON" or "OFF" or "One" or "Zero." Binary is the way digital computers function because they can only represent things as "ON" or "OFF." This binary system contrasts with the "normal" way we write numbers -- i.e. decimal. In decimal, every time you push the number one position to the left, it means you increase it by ten. For example, 100 is ten times the number 10. Computers don't work this way. They work with binary notation. Every time you push the number one position to the right it means you double it. In binary, only two digits are used -- the "0" (zero) and the "1" (the one). If you write the number 10101 in binary, and you want to figure it in decimal as we know it, here's how you do it. 1 is one thing; Zero x 2 = zero; 1 times 2 x 2 = 4; 0 x 2 x 2 x 2 = 0; 1 x 2 x 2 x 2 x 2 = 16. Therefore the total 10101 in binary = 1 + 0 + 4 + 0 + 16 = 21 in decimal.

Binary notation differs slightly from notation used in ASCII or EBCDIC. In ASCII and EBCDIC, the binary values are used for coding of individual characters or keys or symbols on keyboards or in computers. So each string of seven (as in ASCII) or eight (as in EBCDIC) ones and zeros is a unique value -- but not a mathematical one.

ASCII uses a seven bit coding scheme. Thus, the maximum number of different things you can code using seven bits is 128, i.e. 2 x 2 x 2 x 2 x 2 x 2 x 2 = 128. The maximum number represented by a byte (8 bits) or the

IBM EBCDIC coding system is 256. i.e. 2 x 2 x 2 x 2 x 2 x 2 x 2 x 2 = 256. See also BINARY CODE.

BINARY CODE
A code in which every element has only one of two possible values, which may be the presence or absence of a pulse, a pulse, a one or a zero, or high or a low condition for a voltage or current.

BLIND DIALING
Dialing after a pre-set period of time without actually verifying that dial tone is present on the line.

BPS
Bits Per Second. A measure of the speed of data communications. There are many ways to measure bits per second. So don't assume that just because one LAN or other data communications system has a faster bits per seconds, it will transmit your information faster. You have to factor in speed of writing and reading from the disk and the accuracy of transmission. All data communications schemes have error-checking systems, some better than others. Typically such systems force a re-transmission of data if a mistake is detected. You might have a fast, but "dirty" (i.e. lots of errors) transmission medium, which may need lots of re-transmissions. Thus, the "effective" bps of data communications network may actually be quite low. See also BAUD RATE.

CADENCE

In voice processing, cadence is used to refer to the pattern of tones and silence intervals generated by a given audio signal. Examples are busy and ringing tones. A typical cadence pattern is the US ringing tone, which is one second of tone followed by three seconds of silence. Some other countries, such as the UK, use a double ring, which is two short tones within about a second, followed by a little over two seconds of silence.

CALL PROGRESS ANALYSIS

As the call progresses several things happen. Someone dial or touchtones digits. The phone rings. There might be a busy or operator intercept. An answering machine may answer. A fax machine may answer. Call progress analysis is figuring out which is occurring as the call progresses. This analysis is critical if you're trying to build an automated system, like an interactive voice response system.

CALLED STATION IDENTIFICATION

CED. CallED station identification. A 2100 Hz tone with which a fax machine answers a call. See CNG.

CALLED SUBSCRIBER IDENTIFICATION

CSI. This is an identifier whose coding format contains a number, usually a phone number from the remote terminal used in fax.

CALLER ID

Your phone rings. A name pops upon on your phone's screen. It's the name of the person calling you. Or it may be just the caller's phone number. It's called Caller ID and the information about name and/or calling phone number is passed to your phone by your telephone company's central switch. There are basically two forms of "caller ID" -- one provided by your local phone company and one provided by your long distance company (chiefly on 800 calls). Caller ID, generic term, is a term most commonly applied to the service your local phone company provides, usually called CLASS. In CLASS, the information about who's calling and/or their phone number is passed to your phone between the first and second ring signaling an incoming call. See also CLASS and ANI.

CALLING TONE

CNG. CalliNG tone. Also called Auto Fax Tone. This tone is the sound produced by virtually all fax machines when they dial another fax machine. CNG is a medium pitch tone (1100 Hz) that lasts 1/2 second and repeats every 3 ½ seconds. A fax machine will produce CNG for about 45 seconds after it dials. The CNG tone is useful for owners of fax/phone/modem switches. Such switches answer an incoming call. If they hear a CNG tone, they will transfer the call to a fax machine. If they don't, they'll transfer the call to a phone, answering machine or perhaps a modem. Depends on how they're set up. Some fax machines do not transmit a CNG tone with manually-dialed transmissions -- i.e. where the caller picked up the handset on the fax machine, dialed and waited for high-pitched squeal before pushing his fax machine's "start" button. A manual dialed fax transmission will "fool" fax/voice switches. See CED and FACSIMILE

418

CAS

1.Communicating Applications Specification. A high-level API (Application Programming Interface) developed by Intel and DCA that was introduced in 1988 to define a standard software API for fax modems. CAS enables software developers to integrate fax capability and other communication functions into their applications. 2. Channel Associated Signaling.

CCITT

Comite Consultatif Internationale de Telegraphique et Telephonique, which, in English, means the Consultative Committee on International Telegraphy and Telephony. The CCITT is one of the four permanent parts of the International Telecommunications Union, the ITU, based in Geneva Switzerland. The scope of its work is now much broader than just telegraphy and telephony. It now also includes telematics, data, new services, systems and networks (like ISDN). The ITU is a United Nations Agency and all UN members may also belong to the ITU (at present 182), represented by their governments. In most cases the governments give their rights on their national telecom standards to their telecommunications administrations (PITs). But other national bodies (in the US, for example, the State Department) may additionally authorize Recognized Private Operating Agencies (RPOAs) to participate in the work of the CCITT. After approval from their relevant national governmental body, manufacturers and scientific organizations may also be admitted, as well as other international organizations. This means, says the ITU, that participants are drawn from the broad arena. The activities of the CCITT divide into three areas:

Study Groups (at present 15) to set up standard ("recommendations") for telecommunications equipment, systems, networks and services.

Plan Committees (World Plan Committee and Regional Plan Committee) for developing general plans for a harmonized evolution of networks and services.

Specialized Autonomous Groups (GAS, at present three) to produce handbooks, strategies and case studies for support mainly of developing countries.

Each of the 15 Study Groups draws up standards for a certain area - for example, Study Group XVIII specializes in digital networks, including ISDN. Members of Study Groups are experts from administrations, RPOAs, manufacturing companies, scientific or other international organizations - at times there are as many as 500 to 600 delegates per Study Group. They develop standards which have to be agreed upon by consensus. This, says the ITU, can sometimes be rather time-consuming, yet it is a democratic process, permitting active participation from all CCITT member organizations.

The long-standing term for such standards is "CCITT recommendations." As the name implies, recommendations have a non-binding status and they are not treaty obligations. Therefore, everyone is free to use CCITT recommendations without being forced to do so. However, there is increasing awareness of the fact that using such recommendations facilitates interconnection and interoperability in the interest of network providers, manufacturers and customers. This is the reason why CCITT recommendations are now being increasingly applied -- not by force, but because the advantages of standardized equipment are obvious. ISDN is a good example of this.

CCITT has no power of enforcement, except moral persuasion. Sometimes, manufacturers adopt the CCITT specs. Sometimes they don't. Mostly they do. The CCITT standardization process runs in a four-year cycle ending in a Plenary Session. Every four years a series of standards known as Recommendations are published in the form of books. These books are color-coded to represent different four cycles. In 1980 the CCITT published the Orange Books, in 1984 the Red Books and, in 1988, the Blue Books. The CCITT is now more commonly called the ITU, after its parent. See ITU.

The CCITT has now been incorporated into its parent organization, the International Telecommunication Union (ITU). Telecommunication standards are now covered under Telecommunications Standards Sector (TSS). ITU-T (ITU-Telecommunications) replaces CCITT. For example, the Bell 212A standard for 1200 bps communication in North America was referred to as CCITT V.22. It is now referred to as ITU-T V.22.

CED
See CALLED STATION IDENTIFICATION.

CENTRAL OFFICE
Telephone company facility where subscribers' lines are joined to switching equipment for connecting other subscribers to each other, locally and long distance. Also called CO, as in See-Oh. Sometimes the term central

office is the same as the overseas term "public exchange." Sometimes, it means a wire center in which there might be several switching exchanges.

CENTREX

Centrex is a business telephone service offered by a local telephone company from a local central office. Centrex is basically single line telephone service delivered to individual desks (the same as you get at your house) with features, i.e. "bells and whistles," added. Those "bells and whistles" include intercom, call forwarding, call transfer, toll restrict, least cost routing and call hold (on single line phones).

Think about your home phone. You can often get "Custom Calling" features. These features are typically fourfold: Call forwarding, Call Waiting, Call Conferencing and Speed Calling. Centrex is basically Custom Calling, but instead of four features, it has 19 features. Like Custom Calling, Centrex features are provided by the local phone company's central office.

Phone companies peddle Centrex as leased service to businesses as a substitute for that business buying or leasing its own on-premises telephone system – its own PBX, key system or ACD. Before Divestiture in 1984, Centrex was presumed dead. AT&T was, at that time, intent on becoming a major PBX and key system supplier. Then Divestiture came, and the operating phone companies recognized they were no longer part of AT&T, no longer had factories to support, but did have a huge number of Centrex installations providing large monthly revenues. As a result, the local operating companies have injected new life into Centrex, making the service more attractive in features, price, service and attitude. Here are the main reasons businesses go with Centrex as opposed to going with a stand-alone telephone system:

1. Money. Centrex is typically cheaper to get into (the central office already exists). Installation charges can be low. Commitment can also be low, since most Centrex service is leased on a month-to-month basis. So it's perfect for companies planning an early move. There may be some economies of scale, also. Some phone companies are now offering low cost, large size packages.

2. Multiple locations. Companies with multiple locations in the same city are often cheaper with Centrex than with multiple private phone systems and tie lines, or with one private phone system and OPX lines. (An OPX line is an Off Premise eXtension, a line going from a telephone system in one place to a phone in another. It might be used for an extension to the boss's home.)

3. Growth. It's theoretically easier to grow Centrex than a standalone PBX or key system, which usually has a finite limit. With Centrex, because it's provided by a huge central office switch, it's hard, theoretically, to run out of paths, memory, intercom lines, phones, tie lines, CO lines, etc. The limit on the growth of a Centrex is your central office, which may be many thousands of lines.

4. Footprint Space Savings. You don't have to put any switching equipment in your office. All Centrex switching equipment is at the central office. All you need at your office are phones.

5. Fewer Operators because of Centrex's DID features. Fewer operator positions saves money on people and space.

6. Give better service to your customers. With Centrex, each person has their own direct inward dial number. Many people prefer to dial whomever they want directly rather than going through a central operator. Saves time.

7. Better Reliability. When was the last time a central office crashed? Here are some of the features built into modern central offices: redundancy, load-sharing circuitry, power back-up, on-line diagnostics, 24-hour on-site personnel, mirror image architecture, 100% power failure phones, complete DC battery backup and battery power. Engineered to suffer fewer than three hours down time in every 40 years.

8. Non-blocking. Trunking constraints are largely eliminated with Centrex, since a central office is so large.

9. Minimal Service Costs. Repair is cheap. Service time is immediate. People are right next to the machine 24-hours a day. Phones and wires are the only things that require repair on the customers' premises. You can easily plug new phones in, plug them out yourself. All other equipment is in the central office. You need not hold inventory or test equipment.

10. No technological obsolescence. Renting Centrex means a user has the ultimate flexibility -- ability to jump quickly into new technology. Central offices are moving quickly into new technologies, such as ISDN.

11. Ability to manage it yourself. You can now get two important features previously available only on privately-owned self-contained phone systems (like PBXs): 1. The ability for you, the user, to make changes to the programming of your own Centrex installation without having to personally call a phone company representative. 2. The ability to get call detail accounting by extension and then have reports printed by a computer in your office. The phone company does this call accounting by installing a separate data line which carries Centrex call records back to the customer as those calls are made.

The above arguments are pro-Centrex. There are also anti-Centrex arguments. And there's plenty of evidence to argue exactly the opposite. For example, central offices often run out of capacity. The "big" key to Centrex traditionally comes down to price. And, in fact, in some cities the price of Centrex lines is lower than "normal" PBX lines. Of course, you can buy Centrex lines and attach your own PBX or key system to those Centrex lines. The big disadvantage of Centrex is that there are no specialized Centrex phones able to take better advantage of Centrex central office features than normal electronic phones can.

Centrex is known by many names among operating phone companies, including Centron and Cenpac. Centrex comes in two variations -- CO and CU. CO means the Centrex service is provided by the Central Office. CU means the central office is on the customer's premises.

CHANNEL SERVICE UNIT
CSU. A device used to connect a digital phone line (T-1 or Switched 56 line) coming in from the phone company to

either a multiplexer, channel bank or directly to another device producing a digital signal, e.g. a digital PBX, a PC, or data communications device. A CSU performs certain line-conditioning, and equalization functions, and responds to loopback commands sent from the central office. A CSU regenerates digital signals. It monitors them for problems. And it provides a way of testing your digital circuit. You can buy your own CSU or rent one from your local or long distance phone company.

CLASS
Custom Local Area Signaling Services. It is based on the availability of channel interoffice signaling. Class consists of number-translation services, such as call-forwarding and caller identification, available within a local exchange of Local Access and Transport Area (LATA). CLASS is a service mark of Bellcore. Some of the phone services which Bellcore promotes for CLASS are Automatic Callback, Automatic Recall, Calling Number Delivery, Customer Originated Trace, Distinctive Ringing/Call Waiting, Selective Call Forwarding and Selective Call Rejection.

CLIP
Calling Line Identification Presentation.

CNG
See CALLING TONE.

CO
See CENTRAL OFFICE

426

CODEC

Originally CODEC stood for CODer-DECoder, i.e. microprocessor chip. Now the PC industry thinks it stands for COmpression/DEcompression, i.e. an overall term for the technology used in digital video and stereo audio. The original CODEC (still in big use in today's telephony industry) converts voice signals from their analog form to digital signals acceptable to modern digital PBXs and digital transmission systems. It then converts those digital signals back to analog so that you may hear and understand what the other person is saying. In some phone systems, the CODEC is in the PBX and shared by many analog phone extensions. In other phone systems, the CODEC is actually in the phone. Thus the phone itself sends out a digital signal and can, as a result, be more easily designed to accept a digital RS-232-C signal.

COMPANDING

The word is a contraction of the words "compressing" and "expanding." Companding is the process of compressing the amplitude range of a signal for economical transmission and then expanding them back to their original form at the receiving end.

COMPELLED SIGNALING

A signaling method in which the transmission of each signal in the forward direction is inhibited until an acknowledgment of the satisfactory receipt of the previous signal has been sent back from the receiver terminal.

COMPRESSION

Reducing the representation of the information, but not the information itself. Reducing the bandwidth or number of bits needed to encode information or encode a signal, typically by eliminating long strings of identical bits or bits that do not change in successive sampling intervals (e.g., video frames). Compression saves transmission time or capacity. It also saves storage space on storage devices such as hard disks, tape drives and floppy disks.

COMPUTER TELEPHONY

A term coined by Harry Newton to describe the industry that concerns itself with applying computer intelligence to telecommunications devices, especially switches and phones. Computer Telephony has two basic goals: to please customers and to enhance corporate productivity.

CONTINUOUS RECOGNITION

Speech recognition that requires no pause between utterances.

CONTINUOUSLY VARIABLE SLOPE DELTA MODULATION

CVSD. A method for coding analog voice signals into digital signals that uses 16,000 to 64,000 bps bandwidth, depending on the sampling rate.

CPE

See CUSTOMER PREMISE EQUIPMENT.

CROSSTALK

Crosstalk occurs when you can hear someone you did not call talking on your telephone line to another person you did not call. You may also only hear half the other conversation. Just one person speaking. There are several technical causes for crosstalk. They relate to wire placement, shielding and transmission techniques.

CSI

See CALLED SUBSCRIBER IDENTIFICATION.

CSU

See CHANNEL SERVICE UNIT.

CT

See COMPUTER TELEPHONY

CUSTOMER PREMISE EQUIPMENT

CPE. Originally it referred to equipment on the customer's premises which had been bought from a vendor who was not the local phone company. Now it simply refers to telephone equipment -- key systems, PBXs, answering machines, etc. – which reside on the customer's premises. "Premises" might be anything from an office to a factory to a home.

CUT THROUGH

Cut-through is a voice processing term. It's what stops voice prompt playback when a key is pressed. Some of the speech recognition solutions also add cut-through that

429

will stop voice prompt playback as soon as you start talking.

CVSD

See CONTINUOUSLY VARIABLE SLOPE DELTA MODULATION.

D CHANNEL

In an ISDN interface, the "D" channel (the Data channel) is used to carry control signals and customer call data in a packet switched mode. In the BRI (Basic Rate Interface, i.e. the lowest ISDN service) the "D" channel operates at 16,000 bits per second, part of which will carry setup, teardown, ANI and other characteristics of the call. 9,600 bps will be free for a separate conversation by the user. In the PRI (Primary Rate Interface, i.e. ISDN equivalent of T-1), the "D" channel runs at 64,000 bits per second. The D channel provides the signaling information for each of the 23 voice channels (referred to as "B channels"). The actual data which travels on the D channel is much like that of a common serial port. Bytes are loaded from the network and shifted out to the customer site in a serial bit stream. The customer site of course responds with its serial bit stream, too. An example of a data packet sent from the network to indicate a new call has the following components:

-- Customer Site ID
-- Type of Channel Required (Usually a B channel)
-- Call Handle (Not unlike a file handle)
-- ANI and DNIS information
-- Channel Number Requested
-- A Request for a Response

This packet is responded to by the customer site with a format similar to:

-- Network ID
-- Channel Type is OK
-- Call Handle

The packets change as the state of the call changes, and finally ends with one side or the other sending a disconnect notice. The important concept here is the fact the information on the D channel could actually be anything -- any kind of serial data. See also ISDN.

DAC
Digital to Analog Converter. A device which converts digital pulses, i.e. data, into analog signals so that the signal can be used by analog device such as amplifier, speaker, phone, or meter.

DATABASE MANAGEMENT SYSTEM
DBMS. Computer software used to create, store, retrieve, change, manipulate, sort, format and print the information in a database. Database management systems are probably the fastest growing part of the computer industry. Increasingly, databases are being organized so they can be accessible from places remote to the computer they're kept on. The "classic" database management system is probably an airline reservation system.

DB-9
This is the standard nine-pin RS-232-C serial port on the IBM AT and most laptop computers. The term DB-9 is

431

used to describe both the male and female plug. So be careful when you order.

DB-25
The standard 25-pin connector used for RS-232-C serial data communications. In a DB-25 there are 25-pins, with 13 pins in one row and 12 in the other row. DB-25 is used to describe both the male and female plug. So be careful when you order.

DBM
Decibels below 1mW. This should be written as dBm. Output power of a signal referenced to an input signal of 1mW (Milliwatt). Similarly, dBm0 refers to output power, expressed in dBm, with no input signal. (O dBM = 1 milliwatt and -30 dBm = 0.001 milliwatt).

DBMS
See DATABASE MANAGEMENT SYSTEM

DCE
Data Communications Equipment. In the RS-232-C "standard" developed by the Electronic Industries Association, there are DCE devices (typically modems or printers) and DTE (Data Terminal Equipment) devices, which are typically personal computers or data terminals. The main difference between a DCE and DTE is the wiring of pins two and three. But there is, of course, no standardization. When wiring one RS-232-C device to another, it's good to know which device is wired as a DCE and which as a DTE. But it's actually best to go straight to the wiring diagram in the appendix of the device's instruction manual. Then you compare the wiring

diagram of the device you want to connect and build yourself a cable that takes into account the peculiar (i.e. strange) vagaries of the engineers who designed each product. In short, with an RS-232-C connection, the modem is usually regarded as DCE, while the user device (terminal or computer) is DTE. In a X.25 connection, the network access and packet switching node is viewed as the DCE. DCE devices typically transmit on pin 3 and receive on pin 2. DTE (Data Terminal Equipment) devices typically transmit on pin 2 and receive on pin 3. See also DTE.

DCX
Multi-page variation of PCX file format. See PCX.

DELTA MODULATION
A method for converting analog voice to digital form for transmission. It is the second most common method of digitizing voice after Pulse Code Modulation, PCM. Sampling is done in all conversion of analog voice to digital signals. The method of sampling is what distinguishes the various methods of digitization (Delta vs.' PCM, etc.). In delta modulation, the voice signal is scanned 32,000 times a second, and a reading is taken to see if the latest value is greater or less than it was at the previous scan. If it's greater, a "1" is sent. If it's smaller, a "0" is sent.

Delta modulation's sampling rate of 32,000 times a second is four times faster than PCM. But Delta records its samples as a zero (0) or a one (1), while PCM takes an 8-bit sample. Thus PCM encodes voice into 64,000 bits per second, while Delta codes it into 32,000. Because delta has fewer bits, it could theoretically produce a poorer representation of the voice. In actual fact, the human ear can't hear the difference between a PCM and a Delta encoded voice conversation.

Delta modulation has much to recommend it, especially its use of fewer bits. Unfortunately no two delta modulation schemes are compatible with each other. So to get one delta-mod digital PBX to speak to another, you have to convert the voice signals back to analog. With AT&T making T-1 a de facto digital encoding scheme, PCM has become the de facto standard for digitally encoding voice. And although there are three types of PCM in general use, they can be made compatible on direct digital basis (i.e. without having to go back to analog voice). One problem with PCM is that American manufacturers typically put twenty four 64,000 bit per second voice conversations on a channel and call it T-1. The Europeans put 30 conversations on their equivalent transmission path. Thus, you can't directly interface the American and the European systems. But there are "black boxes" available...(In this business, there are always black boxes available.)

DEMARCATION POINT
The point of a demarcation and/or interconnection between telephone company communications facilities and terminal equipment, protective apparatus, or wiring at a subscriber's premises. Carrier-installed facilities at or constituting the demarcation point consist of a wire or a jack conforming to Subpart F of Part 68 of the FCC Rules.

DIALED NUMBER IDENTIFICATION SERVICE
DNIS is a feature of 800 and 900 lines that provides the number the caller dialed to reach the attached computer telephony system (manual or automatic). Using DNIS capabilities, one trunk group can be used to serve multiple applications. The DNIS number can be provided in a number of ways, inband or out-of-band, ISDN or via a separate data channel. Generally, a DNIS number will be used to identify to the answering computer telephony system the "application" the caller dialed. For example, a 401K status program may be offered by a service provider to a number of different companies. The employees of each company are provided their own 800 number to call to access their account status. When the computer telephony system sees the incoming DNIS number, it will know to which company the call was directed, and can so answer the phone correctly with a customized "you have reached the 401K line for xyz company. Please enter your personal account code and password..."

Here's another application: You use one 800 phone number for testing your advertisements on TV stations in Phoenix; another number for testing your ads on TV stations in Chicago; and yet another for Milwaukee. The DNIS information can be used in a multitude of ways -- from playing different messages to different people, to routing those people to different operators, to routing those people to the same operators, but flashing different messages on their screens, so the operators answer the phone differently. In Ireland, incoming toll free phone calls from the rest of Europe arrive with DNIS. As a result a phone call arriving from Germany is routed to a computer telephony system playing messages in German. The advantage of DNIS is basically economic: You simply need fewer phone lines. Without DNIS you would need at

435

least one phone line for every different 800 or 900 number you gave out to your callers. Make sure you understand the difference between DNIS and ANI and Caller ID. DNIS tells you the number your caller called. ANI or Caller ID is the number your caller called from.

DIAL TONE
The sound you hear when you pick up a telephone. Dial tone is a signal (350 + 440 Hz) from your local telephone company that it is alive and ready to receive the number you dial. If you have a PBX, dial tone will typically be provided by the PBX. Dial tone does not come from God or the telephone instrument on your desk. It comes from the switch to which your phone is connected to.

DID
See DIRECT INWARD DIALING.

DIGITAL SIGNAL
A discontinuous signal. One whose state consists of discrete elements, representing very specific information. When viewed on an oscilloscope, a digital signal is "squared." This compares with an analog signal which typically looks more like a sine wave, i.e. curvey. Usually amplitude is represented at discrete time intervals with a digital value.

DIGITAL SIGNAL PROCESSOR
A specialized digital microprocessor that performs calculations on digitized signals that were originally analog (e.g. voice) and then sends the results on. There

are two main advantages of DSPs -- first, they have powerful mathematical computational abilities, much more than normal computer microprocessors. DSPs need to have heavy mathematical computation skills because manipulating analog signals requires it. For example, DSPs are often called upon to compress video signals. Each sample must be examined and processed. And all done in very little time. The second advantage of a DSP lies in the programmability of digital microprocessors. Just as digital microprocessors have operating systems, so DSPs are now acquiring their very own operating systems. DSPs are used extensively in telecommunications for tasks such as echo cancellation, call progress monitoring, voice processing and for the compression of voice and video signals. They are also used in devices from fetal monitors, to anti-skid brakes, seismic and vibration sensing gadgets, super-sensitive hearing aids, multimedia presentations and low cost desktop fax machines. DSPs are replacing the dedicated chipsets in modems and fax machines with programmable modules -- which, from one minute to another, can become a fax machine, a modem, a teleconferencing device, an answering machine, a voice digitizer and device to store voice on a hard disk, to a proprietary electronic phone. DSPs will do (and are already doing) for the telecom industry what the general purpose microprocessor (e.g. Intel's 80286 or 80386) did for the personal computer industry. DSPs are made by Analog Devices, AT&T, Motorola, NEC and Texas Instruments, among others.

DIRECT INWARD DIALING
DID. You can dial inside a company directly without going through the attendant. This feature used to be an

exclusive feature of Centrex but it can now be provided by virtually all modern PBXs and some modern hybrids, but you must connect via specially configured DID lines from your local central office. A DID (Direct Inward Dial) trunk is a trunk from the Central office which passes the last two to four digits of the Listed Directory Number to the PBX or hybrid phone system, and the digits may then be used verbatim or modified by phone system programming to be the equivalent of an internal extension. Therefore, an external caller may reach an internal extension by dialing a 7-digit central office number. Notice: DID is different from a DIL (Direct-In-Line) where a standard, both-way central office trunk is programmed to always ring a specific extension or hunt group. DID lines cannot be used for outdial operation, since there is no dialtone offered.

DISCONNECT
The breaking or release of a circuit connecting two telephones or data devices.

DISCRETE RECOGNITION
In speech recognition, Discrete Recognition refers to an isolated word. A discrete word is preceded and followed by silence, hence isolated in speech. Discrete words need to be separated by about half a second of silence when spoken to a discrete recognizer.

DIVESTITURE
On January 8, 1982 AT&T signed a Consent Decree with the U.S. Department of Justice, stipulating that on midnight December 30, 1983, AT&T would divest itself of

its 22 telephone operating companies. According to the terms of the Divestiture, those 22 operating Bell telephone companies would be formed into seven regional holding companies of roughly equal size. Terms of the Divestiture placed business restrictions on AT&T and the BOCs. Those restrictions were threefold: The BOCs weren't allowed into long distance, equipment manufacturing, or information services. AT&T wasn't allowed into local telecommunications (i.e. to compete with the BOCs). But it was allowed into computers. The federal Judge overseeing Divestiture, Judge Harold Greene, is slowing the lifting the restrictions against the BOCs being allowed into information services. He has stayed firm on the other two -- equipment manufacturing and long distance.

DLL
Dynamic Link Library. A feature of OS/2 and Windows that allow executable code modules to be loaded on demand and linked at run time. This lets library code be field-updated -- transparent to applications – and then unloaded when they are no longer needed.

DM
See DELTA MODULATION.

DNIS
See DIALED NUMBER IDENTIFICATION SERVICE.

DSP
See DIGITAL SIGNAL PROCESSOR

DTE
Data Terminal Equipment. In the RS-232-C standard specification, the RS-232-C is connected between the DCE (Data Communications Equipment) and a DTE. The main difference between a DCE and a DTE is that pins two and three are reversed. See also DCE.

DTMF
Dual Tone Multi Frequency. A fancy term describing push button or Touchtone dialing. (Touchtone is a not registered trademark of AT&T, though until 1984 it was.) In DTMF, when you touch a button on a push button pad, it makes a tone, actually a combination of two tones, one high frequency and one low frequency. Thus the name Dual Tone Multi Frequency. In U.S. telephony, there are actually two types of "tone" signaling, one used on normal business or home push button/touchtone phones, and one used for signaling within the telephone network itself. When you go into a central office, look for the test board. There you'll see what looks like a standard touchtone pad. Next to the pad there'll be a small toggle switch thatallows you to choose the sounds the touchtone pad will make – either normal touchtone dialing (DTMF) or the network version (MF).

The eight possible tones that comprise the DTMF signaling system were specially selected to easily pass through the telephone network without attenuation and with minimum interaction with each other. Since these tones fall within the frequency range of the human voice, additional considerations were added to prevent the human voice from inadvertently imitating or "falsing" DTMF signaling digits. One way this was done to break the tones into two groups, a high frequency group and a low frequency group. A valid DTMF tone has only one tone in each group. Here is a table of the DTMF digits with their respective frequencies. One Hertz (abbreviated Hz.) is one cycle per second of frequency.

Digit	Low frequency	High frequency
1	697 Hz.	1209 Hz.
2	697	1336
3	697	1477
4	770	1209
5	770	1336
6	770	1477
7	852	1209
8	852	1336
9	852	1477
0	941	1336
*	941	1209
#	941	1477

There are four other digits defined in the DTMF system and usable for specialized applications that cannot be generated by standard telephones. They are:

A	697 Hz.	1633 Hz.
B	770	1633
C	852	1633
D	941	1633

Normal telephones (yours and mine) have 12 buttons, thus 12 combinations. Government Autovon (Automatic

Voice Network) telephones have 16 combinations, the extra four (those above) being used for "precedence," which in Federal government parlance is a designation assigned to a phone call by the caller to indicate to communications personnel the relative urgency (therefore the order of handling) of the call and to the called person the order in which the message is to be noted.

E-1
The European equivalent of the North American 1.544 Mbps T-1,except that E-1 carries information at the rate of 2.048 megabits per second. This is the rate used by European CEPT carriers to transmit 30 64 Kbps digital channels for voice or data calls, plus a 64 Kbps channel for signaling, and a 64 Kbps channel for framing (synchronization) and maintenance. CEPT stands for the Conference of European Postal and Telecommunication Administrations. Since robbed-bit signaling is not used (as it is for T-1 in North America) all 8 bits per channel are used to code the waveshape sample. See T-1 and Earth Recall.

EASE
Computer Telephony application generator software from Expert Systems, Inc.

EBCDIC
(Pronounced Eb-si-dick.) Extended Binary Coded Decimal Interexchange Code. It is the way IBM codes characters, letters and numbers into a digital binary stream for use in its larger computers. EBCDIC codes characters into

eight bits. This gives it 256 possible characters, 2 x 2 x 2 x 2 x 2 x 2 x 2 x 2 = 256.

EBCDIC is mainly used in IBM mainframes and minicomputers, while ASCII is used in IBM and non-IBM desktop microcomputers. EBCDIC is not compatible with ASCII, meaning that a computer which understands EBCDIC will not understand ASCII. But there are many real-time and non-real time translation programs that will convert text files back and forth. A good program is Word-For-Word from MasterSoft. See ASCII.

ECM
See ERROR CORRECTION MODE.

EIA
Electronic Industries Association. A Washington, D.C. trade organization of manufacturers which sets standards for use of its member companies, conducts educational programs and lobbies in Washington for its members' collective prosperity. In April 1988, a new association called the Telecommunications Industry Association was formed by a merger of the US Telecommunications Suppliers Association (USTSA) and the Electronics Industries Association's Information and Telecommunications Technologies Group (EIA/ITG).

ENCODING
The process of converting data into code or analog voice into a digital signal. See also PCM and ADPCM.

ERROR CORRECTION MODE
ECM. An enhancement to Group 3 fax machines. Encapsulated data within HDLC frames providing the received with an opportunity to check for, and request retransmission of garbled data. See FACSIMILE and V.17.

EURO-ISDN
The European implementation of ISDN. In some ways, it differs from North American National ISDN-1. Users, can however, call from the United States to Europe and complete ISDN calls.

It is uncertain (translate highly improbable) whether they can carry their end-user equipment from the United States and use it effectively in Europe.

FALSING
In telecom signaling, DTMF tones are created using specific combinations of frequencies to prevent the possibility of "falsing." Falsing is the condition where a DTMF detector incorrectly believes a DTMF is present when in fact it is actually a combination of voice, noise and/or music.

FOURIER'S THEOREM
In the early 1800s, the French mathematician Emile Fourier proved that a repeating, time-varying function may be expressed as the sum of a (possibly infinite) series of sine and cosine waves.

Digital data is a bit stream, which can be sent as a sequence of square waves. Fourier's Theorem shows that to send a square wave (digital signal), a series of sine waves (analog signals) are actually summed together. If 1,000 square waves are to be sent every second, for example, the frequency components of the sine waves that are summed together are 1 kHz, 3 kHz, 5 kHz, 7 kHz, etc. The point of this analysis is to show that high frequency signals are required to form a stable, recognizable square wave. As the bit rate increases, the square wave frequency increases and the width of the square waves decrease. Thus, narrower square waves require sine waves of even higher frequencies to form the digital signal.

FRAME

1. Generally, a group of data bits in a specific format, with a flag at each end to indicate the beginning and end of the frame. The defined format enables network equipment to recognize the meaning and purpose of specific bits. The group of bits are sent serially (one after another). Generally a frame is a logical transmission unit. A frame usually contains its own control information for addressing and error checking. A frame is the basic data transmission unit employed in bit-oriented protocols. In this way, a frame is similar to a block.

2. One complete cycle of events in time division multiplexing. The frame usually includes a sequence of time slots for the various sub channels as well as extra bits for control, calibration, etc.

FRAMING BIT
A bit used for frame synchronization purposes. A bit at a specific interval in a bit stream used in determining the beginning or end of a frame. Framing bits are non-information-carrying bits used to make possible the separation of characters in a bit stream into lines, paragraphs, pages, channels etc. Framing in a digital signal is usually repetitive.

FREQUENCY
The rate at which an electrical current alternates, usually measured in Hertz. Hertz is a unit of measure which means "cycles per second." So, frequency equals the number of complete cycles of current occurring in one second.

GAIN
The increase in signaling power that occurs as the signal is boosted by an electronic device. It's measured in decibels (dB).

GLARE
Glare occurs when both ends of a telephone line or trunk are seized at the same time for different purposes or by different users.

HDLC
High level Data Link Control. A standard bit-oriented protocol developed by the International Standards Organization (ISO). In HDLC, control information is always placed in the same position. And specific bit

patterns used for control differ dramatically from those used in representing data, so that errors are less likely to occur. SDLC and ADCCP are similar protocols.

HLLAPI
High Level Language Applications Programming Interface. An IBM API.

HOOKSWITCH
Also called SWITCHHOOK or switch hook. The place on your telephone instrument where you lay your handset. A hookswitch was originally an electrical "switch" connected to the "hook" on which the handset (or receiver) was placed when the telephone was not in use. The hookswitch is now the little plunger at the top of most telephones which is pushed down when the handset is resting in its cradle (on-hook). When the handset is raised, the plunger pops up and the phone goes off-hook. Momentarily depressing the hookswitch (up to 0.8 of a second) can signal various services such as calling the attendant, conferencing or transferring calls.

HUNT GROUP
A series of telephone lines organized in such a way that if the first line is busy the next line is hunted and so on until a free line is found. Often this arrangement is used on a group of incoming lines. Hunt groups may start with one trunk and hunt downwards. They may start randomly and hunt in clockwise circles. They may start randomly and hunt in counter-clockwise circles. Inter-Tel uses the terms "Linear, Distributed and Terminal" to refer to different types of hunt groups. In data communications, a

447

hunt group is a set of links which provides a common resource and which is assigned a single hunt group designation. A user requesting that designation may then be connected to any member of the hunt group. Hunt group members may also receive calls by station address.

HYBRID PBX/KEY SYSTEM

Term used to describe a system which has attributes of both Key Telephone Systems and PBXs. The one distinguishing feature these days is that a hybrid key system can use normal single line phones in addition to the normal electronic key phones. A single line phone behind a hybrid works very much like a single line phone behind a PBX. The second distinguishing feature of a hybrid is that it's "non-squared." This means that not every trunk appears as a button on every phone in the system -- as occurs on virtually every electronic key system manufactured today.

Hz

Abbreviation for Hertz. A measurement of frequency in cycles per second. A hertz is one cycle per second.

ISDN

Integrated Services Digital Network. ISDN comes today in two basic flavors -- BRI, which is 144,000 bits per second and designed for the desktop, and PRI which is 1,544,000 bits per second and designed for telephone switches, computer telephony and voice processing systems. Neither ISDN BRI or ISDN PRI is a standard service, though there are several "standard" configurations. ISDN BRI is a wonderful service in your home or office because it can give you videoconferencing, and ultrafaster data communications. But it is not an easy service to get up and running. The best advice I can give you is: 1. Figure out what you want to do with your ISDN. 2. Find which equipment you're going to need that will do the best job for you. 3. Call the manufacturer of that equipment, tell him where you're located and ask him which ISDN service to order. 4. After he tells you, order your ISDN service from your local phone company. 5. Then buy the equipment. 6. Allow yourself at least a month to get up and running. 7. Any ISDN equipment you install in a PC will cause major interrupt problems. Make sure you know which interrupts your PC is using for what.

ISDN is essentially a totally new concept of what the world's telephone system should be. According to AT&T, today's public switched phone network has the following limitations: 1. Each voice line is only 4 KHz, which is very narrow, which limits also the speed you can send data across. 2. Most signaling is in-band signaling, which is very consuming of bandwidth (i.e. it's expensive and inefficient). 3. The little out-of-band signaling that exists today runs on lines separate to the network. This includes signaling for PBX attendants, hotel/motel, Centrex and PBX calling information. 4. Most users have separate voice and data networks, which is inefficient, expensive and limiting. 5. Premises telephone and data equipment must be separately administered from the network it runs on. 6. There is a wide and growing variety of voice, data and digital interface standards, many of which are incompatible.

ISDN's "vision" is to overcome these deficiencies in four ways: 1. By providing an internationally accepted standard for voice, data and signaling. That standard has pretty well achieved, though don't try and take North American ISDN equipment to Europe. 2. By making all transmission circuits end-to-end digital. 3. By adopting a standard out-of-band signaling system. 4. By bringing significantly more bandwidth to the desktop.

One of the best features of ISDN is the speed of dialing. Instead of 20 seconds for a call to go through on today's old analog network, with ISDN it takes less than a second. It's beautiful. Here are some sample ISDN services:

Call waiting: A line is busy. A call comes in. The user knows who is calling. He can then accept, reject, ignore, transfer the call.

Credit card calling: Automatic billing of certain or all calls into accounts independent of the calling line/s.

Calling line identification presentation: Provides the calling party the ISDN "phone" number, possibly with additional address information, of the called party. Such information may flash across the screen of an ISDN phone or be announced by a synthesized voice. The called party can then accept, reject or transfer the call. If the called party is not there, then his/her phone will automatically record the incoming call's phone number and allow automatic callbacks when he/she returns or calls back in from elsewhere.

Calling line identification restriction: Restricts presentation of the calling party's ISDN "phone" number, possibly with additional address information, to the called party.

450

Closed user group: Restricts conversations to or among a select group of phone numbers, local, long distance or international.

Collaborative Computing. Work on the same document or drawing or design with someone 10,000 miles away. With ISDN, it doesn't really matter where members of the design team live.

Simultaneous Data Calls: Two users can talk and exchange information over the D packet and/or the B circuit or packet switched channel.

There are two major problems to the widespread acceptance of ISDN: First, the cost of ISDN terminal equipment is too high. Second, the cost of upgrading central office hardware and software to ISDN is too high. Both costs are coming down. In early, 1995 Pacific Bell announced that it would install one million lines of ISDN BRI by the end of 1998. And, to make this happen, it dropped its ISDN monthly prices to an affordable $24.95 a month for residences and $26.50 a month for businesses.

There are three basic configurations you can get ISDN:

1. The 2B+D "S" interface (also called the "T" interface). The 2B+D is called the Basic Rate Interface (BRI). The "S" interface uses four unshielded normal telephone wires (two twisted wire pairs) to deliver two "Bearer" 64,000 bits per second channels and one "data" signaling channel of 16,000 bits per second. An S-interfaced phone can be located up to one kilometer from the central office switch driving it. Each of the two 64 kpbs "bearer" or B channels can be used to carry a voice conversation, or one high

speed data or several data channels, which are multiplexed into zone 64 kbps high speed data line. The "D" channel of 16 kbps will carry control and signaling information to set up and break down the voice and data calls. The "D" channel can also carry data up to 9600 bits per second in addition to the control and signaling information. Signaling and control on the D channel conforms to a protocol (LAPD) and a messaging structure (Q.931). These two allow intelligent endpoints and switching nodes from different vendors to talk a common language and thus be able to transfer features across a network, from one switch to another, e.g. to transfer a Centrex call across town through several switches and to have it arrive at the end phone with the calling party's name.

2. The 2B+D "U" interface. This "U" interface delivers the same two 64 kbps bearer channels and one 16 kbps data channel, except that it uses 2-wires (one pair) and can work at 5-10 kilometers from the central office switch driving it. The "U" interface is the most common ISDN interface. It carries 160,000 bits per second from the central office to your home or office. Of those 160,000 bits, two are used for 64,000 bps Bearer (B) channels and one is used by the subscriber for 16,000 bps of data (the D channel). The other 16,000 bps is used by the network for signaling between the black box on the subscriber premises and the central office. The idea is to get the ISDN "U" interface working to 18,000 feet -- the average length of a North American subscriber local loop. You connect the two "U" wires (local loop pair) coming in from your local ISDN CO into a black box about the size of desk printing calculator, called an NT-1. Out the side of the black box comes four wires, which are called the "S Bus." Onto these four wires you can attach, in a loop

452

configuration (also called single bus), as many as eight ISDN terminals -- telephones, fax machines, etc.

3. The 23B+D or 30B+D. This is called the Primary Rate Interface (PRI). At 23B+D, it is 1.544 megabits per second. At 30B+D, it is 2.048 megabits per second. The first, 23B+D is the standard T-1 line in the U.S. which operates on two pairs. The second 30B+D is the standard T-1 line in Europe, which also operates on two pairs.

Integral to ISDN's ability to produce new customer services is CCITT Signaling System 7. This is a ITU-T recommendation which does two basic things: First, it removes all phone signaling from the present network onto a separate packet switched data network, thus providing enormous economies of bandwidth. Second, it broadens the information that is generated by a call, or call attempt. This information -- like the phone number of the person who's calling -- will significantly broaden the number of useful new services the ISDN telephone network of tomorrow will be able to deliver.

ISDN has "enjoyed" many "meanings," including I Still Don't Know and It Still Does Nothing to, its most recent, I Smell Dollars Now.

ITU-T
Intl Telecommunication Union-Telecommunication sector. The organization which has replaced the CCITT as the world's leading telecommunications standards organization. The ITU-T is the international organization that defines standards for telegraphic and telephone equipment. For example, the Bell 212A standard for 1200 bps communication in North America is observed

internationally as ITU-T V.22. For 2400 bps communication, most U.S. manufacturers observe V.22 bis, etc. See V.XX.

IVR

Interactive Voice Response. Think of IVR as a voice computer. Where a computer has a keyboard for entering information, an IVR uses remote touchtone telephones. Where a computer has a screen for showing the results, an IVR uses a prerecorded human voice that is stored (digitized) on a hard drive, in addition it can use a synthesized voice (computerized voice) for read back information that is constantly changing. (The synthesized voice is commonly referred to as Text-to-Speech.

Whatever a computer can do, an IVR can too -- from looking up train timetables to moving calls around an automatic call distributor (ACD). The only limitation on an IVR is that you can't present as many alternatives on a phone as you can on a screen. The caller's brain simply won't remember more than a few. With IVR, you have to present the menus in smaller chunks. See also COMPUTER TELEPHONY.

The benefits of Interactive Voice Response are obvious. By automating the retrieval and processing of information by phone, you can "give data a voice" and "add intelligence to the phone call." By doing that, you can:

Put information to work. The classic IVR "killer app" takes an existing database (e.g., a magazine's article archives, a freight company's package-tracking system) and makes it available by phone (or other media, such as fax, e-mail, or DSVD -- Digital Simultaneous Voice and

454

Data). You can automate telephone-based tasks. From "bank by phone" to "find my package," to "sell me an airline ticket," to "validate my new credit card," IVR gives access to and takes in information; performs record-keeping, and makes sales, 24 hours a day -- supplementing or standing in for human personnel. IVR can add value to communications. Any call-handling operation can profit from IVR. Used as a front-end for an ACD, an IVR system can ask questions (e.g., "what's your product serial code?") that help routing and enable more intelligent and informed call processing (by people or automatic systems). IVR far supersedes more rudimentary technologies (such as Caller ID) in such applications. Used in place of traditional on-hold programming, IVR can add interactive value to what would otherwise be wait-time. The IVR can be used to distribute information, make callers aware of specials -- even provide entertainment. The result: fewer callers drop off queue; you make more sales.

KBPS
Kilo-Bits Per Second or one thousand bits per second.

KEY TELEPHONE SYSTEM
A system in which the telephones have multiple buttons permitting the user to select outgoing or incoming central office phone lines directly. With a key system you don't have to dial 9 to get obtain an outside line. With a PBX, you have to dial "9" to make a call outside the building. Dialing 9 is the major difference these days between a key telephone systems and PBXs. PBXs are usually, but not always, larger.

KHZ

KiloHertz. One thousand hertz or cycles per second. Written kHz.

LOOP CURRENT DETECTION

When a modem, telephone or fax card (etc.) seizes the line (i.e. completes the connection between tip and ring terminals of the telephone cable) current flows from the positive battery supply in the telephone central office, through the twisted pair in the loop, through the card (or phone) and back to the central office negative terminal where it is detected, showing that this telephone or telephone device is off hook. The fax card or modem can detect problems such as disconnects, shutting down the connection or a busy signal.

LOOP START

You "start" (seize) a phone line or trunk by giving it a supervisory signal. That signal is typically taking your phone off hook. There are two ways you can do that -- ground start or loop start. With loop start, you seize a line by bridging through a resistance the tip and ring (both wires) of your telephone line. The Loop Start trunk is the most common type of trunk found in residential installations. The ring lead is connected to -48V and the tip lead is connected to OV (ground). To initiate a call, you form a "loop" ring through the telephone to the tip. Your central office rings a telephone by sending an AC voltage to the ringer within the telephone. When the telephone goes off-hook, the DC loop is formed. The central office detects the loop and the fact that it is drawing DC current and stops sending the ringing voltage. In ground start trunks, ground Starting is a handshaking routine that is performed by the central office and the PBX prior to making a phone call. The central office and the PBX agree to dedicate a path so incoming and outgoing calls cannot conflict, so "glare" cannot occur. See GLARE. Here are two questions that help in understanding:

How does a PBX check to see if a CO Ground Start trunk has been dedicated?

To see if the trunk has been dedicated, the PBX checks to see if the TIP lead is grounded. An undedicated Ground Start Trunk has an open relay between OV (ground) and the TIP lead connected to the PBX. If the trunk has been dedicated the CO will close the relay and ground the TIP lead.

How does a PBX indicate to the CO that it requires the trunk?

A CO ground start trunk is called by the PBX CO Caller circuit. This circuit briefly grounds the ring lead causing DC current to flow. The CO detects the current flow and interprets it as a request for service from the PBX.

MBPS
Mega-Bits Per Second. One million bits per second. Written Mbps or Mbit/s.

MF
Multi-Frequency pulsing. An in-band address signaling method in which ten decimal digits (the numbers on the touchtone pad) and five auxiliary signals are each

represented by selecting two frequencies and combining them into one "musical" sound.. The frequencies are selected from six separate frequencies -- 700, 900, 1100, 1300, 1500 and 1700 Hz.

MHZ
Megahertz. A unit of frequency denoting one million Hz or cycles per second. Written MHz.

MU-LAW
The PCM voice coding and companding standard used in Japan and North America. A PCM encoding algorithm where each digitized sample is represented by eight bits, thus a 64k transmission rate -- the standard rate for encoding voice -- is actually 8K samples/second. A sample consists of a sign bit, a three bit segment specifying a National logarithmic range, and a four bit step offset into the range. All bits of the sample are inverted before transmission. See COMPANDING and PCM.

NYQUIST THEOREM
In communications theory, a formula stating that two samples per cycle is sufficient to characterize an analog signal. In other words, the sampling rate must be twice the highest frequency component of the signal (i.e., sample 4 KHz analog voice channels 8000 times per second.).

ODBC
Open Database Connectivity (ODBC) is Microsoft's strategic interface for accessing data in a heterogeneous

environment of relational and non-relational database management systems. Based on work-in-progress on the Call Level Interface (CLI) specification from the SQL Access Group, ODBC provides a vendor-neutral way of accessing data in a variety of personal computer, minicomputer and mainframe databases.

PAM

Pulse Amplitude Modulation. Process of representing a continuous analog signal (a voice conversation) with a series of discrete analog samples. This concept is based on the information theory which suggests that the signal can be accurately recreated from a sufficient sample. Why bother? Sampling allows several signals to then be combined on a channel that otherwise would only carry one telephone conversation. PAM was used as part of a method of switching phones calls in several PBXs. It is not a truly "digital" switching system. PAM is the basis of PCM, Pulse Code Modulation. See PCM and T-1.

PART 68 REQUIREMENTS

Specifications established by the FCC as the minimum acceptable protection communications equipment must provide the telephone network. Meeting these requirements does not certify that equipment performs any task. Part 68 is the section of Title 47 of the Code of Federal Regulations governing the direct connection of telecommunications equipment and premises wiring with the public switched telephone network and certain private line services, e.g., foreign exchange lines (customer premises end), the station end of off-premises stations associated with PBX and Centrex services, trunk-to-station tie lines (trunk end only), and switched service network station lines (common control switching arrangements); and the direct connection of all PBX (or similar) systems to private line services for tie trunk type interfaces, off-premises station lines, automatic identified outward dialing and message registration. These rules provide the technical, procedural and labeling standards under which direct electrical connection of customer-provided telephone equipment, systems, and protective apparatus may be made to the nationwide network without causing harm and without a requirement for protective circuit arrangements in the service provider's network. Form 730 Application Guide is a collection of literature you'll need to register your telephone/telecom equipment under Part 68 of Title 47 at the Federal Communications Commissions. To get this material (it's free) drop a line or call the Federal Communications Commission, Washington DC 20554.

PBX
Private Branch eXchange. A private (i.e. you, as against the phone company owns it), branch (meaning it is a small phone company central office), exchange (a central office was originally called a public exchange, or simply an exchange). In other words, a PBX is a small version of the phone company's larger central switching office. A PBX is also called a Private Automatic Branch Exchange, though that has now become an obsolete term. In the very old days, you called the operator to make an external call, except in Europe. Then later someone made a phone system that you simply dialed nine (or another digit -- in Europe it's often zero), got a second dial tone and dialed some more digits to dial out, locally or long distance. So, the early name of Private Branch Exchange (which needed an operator) became Private AUTOMATIC

Branch Exchange (which didn't need an operator). Now, all PBXs are automatic. And now they're all called PBXs, except overseas where they still have PBXs that are not automatic.

At the time of the Carterfone decision in the summer of 1968, PBXs were electro-mechanical step-by-step monsters. They were 100% the monopoly of the local phone company. AT&T was the major manufacturer with over 90% of all the PBXs in the U.S. GTE was next. But the Carterfone decision allowed anyone to make and sell a PBX. And the resulting inflow of manufacturers and outflow of innovation caused PBXs to go through five, six or seven generations -- depending on which guru you listen to. Anyway, by the fall of 1991, PBXs were thoroughly digital, very reliable, and very full featured. There wasn't much you couldn't do with them. They had oodles of features. You could combine them and make your company a mini-network. And you could buy electronic phones that made getting to all the features that much easier. Sadly, by the late 1980s the manufacturers seemed to have finished innovating and were into price cutting. As a result, the secondary market in telephone systems was booming. Fortunately, that isn't the end of the story. For some of the manufacturers in the late 1980s figured that if they opened their PBXs' architecture to outside computers, their customers could realize some significant benefits. (You must remember that up until this time, PBXs were one of the last remaining special purpose computers that had totally closed architecture. No one else could program them other than their makers.) Some of the benefits customers could realize from open architecture included:

- Simultaneous voice call and data screen transfer.

- Automated dial-outs from computer databases of phone numbers and automatic transfers to idle operators.

- Transfers to experts based on responses to questions, not on phone numbers.

An alternative to getting a PBX is to subscribe to your local telephone company's Centrex service. For a long explanation on Centrex and its benefits, see CENTREX. Here are some of the benefits of a PBX versus Centrex:

1. Ownership. Once you've paid for it, you own it. There are obvious financial and tax benefits.

2. Flexibility. A PBX is a far more flexible than a central office based Centrex. A PBX has more features. You can change them faster. You can expand faster. Drop another card in, plug some phones in, do your programming and bingo you're live.

3. Centrex benefits. You can always put Centrex lines behind a PBX and get the advantages of both. In some towns, Centrex lines are cheaper than PBX lines. So buy Centrex lines and put them behind your PBX. Make sure you don't pay for Centrex features your PBX already has. (It has most.)

4. PBX phones. There are really no Centrex phones -- other than a few Centrex consoles. If you want to take advantage of Centrex features, you have to punch in cumbersome, difficult-to-remember codes on typically

single line phones. PBXs have electronic phones, often with screens and dedicated buttons. They're usually a lot easier to work. A lot easier to transfer a call. Conference another, etc. A lot more productive.

5. Footprint savings. Modern PBXs take up room, more than Centrex. But the space they take up is far less than it used to be. PBXs are getting smaller.

6. Voice Processing/Automated Attendants. Centrex's DID (Direct Inward Dialing) feature was always pushed as a big "plus." You saved operators. However, you can now do operator-saving things with PC-based voice processing and automated attendants you couldn't do five years ago. These things work better with on-site standalone PBXs than with distant, central office based Centrex. Moreover, virtually every PBX in existence today supports DID. You can dial directly into PBXs and reach someone at their desk just as easily as you can dial directly using Centrex.

7. Open Architecture. Most PBXs have open architecture. Central offices don't.

8. Good Reliability. There have been sufficient central office crashes and sufficient improvement in the reliability of PBXs that you could happily argue that the two are on a par with each other today. Both are equally reliable, or unreliable. The only caveat, of course, is that you back your PBX up with sufficient batteries that it will last a decent power outage. Of course, that assumes that your people will be prepared to hang around and answer the phones during a blackout.

9. Expansion. Central offices are big. Allegedly you can grow your lines to whatever size you want. In contrast,

PBXs have finite growth. It's true about PBXs. But it's equally true about central offices. I've personally heard too many stories about central office line shortages to believe in the nonsense about "infinite Centrex" growth. Fact is central offices grow out, just like PBXs. Given the tight economy of recent years, local phone companies have not been buying the central offices they should have. And they have been filling central offices up a little too tight for my taste.

10. Technological obsolescence. Allegedly central offices are upgraded faster than PBXs and therefore are always up to date technologically. It's nonsense. The life cycle of a typical central office was 40 years until recently. It's now around 20 years. Think of what's happened to PCs in the past 10 years -- the IBM PC debuted only in 1981 -- and you can imagine how obsolete many of the nation's central offices are.

PCM
Pulse Code Modulation. The most common method of encoding an analog voice signal into a digital bit stream. First, the amplitude of the voice conversation is sampled. This is called PAM, Pulse Amplitude Modulation. This PAM sample is then coded (quantized) into a binary (digital) number. This digital number consists of zeros and ones. The voice signal can then be switched, transmitted and stored digitally. There are three basic advantages to PCM voice. They are the three basic advantages of digital switching and transmission. First, it is less expensive to switch and transmit a digital signal. Second, by making an analog voice signal into a digital signal, you can interleave it with other digital signals -- such as those from computers or facsimile machines.

464

Third, a voice signal which is switched and transmitted end-to-end in a digital format will usually come through "cleaner," i.e. have less noise, than one transmitted and switched in analog. The reason is simple: An electrical signal loses strength over a distance. It must then be amplified. In analog transmission, everything is amplified, including the noise and static the signal has collected along the way. In digital transmission, the signal is "regenerated," i.e. put back together again, by comparing the incoming signal to a logical question: Is it a one or a zero? Then, the signal is regenerated, amplified and sent along its way.

PCM refers to a technique of digitization. It does not refer to a universally accepted standard of digitizing voice. The most common PCM method is to sample a voice conversation at 8000 times a seconds. The theory is that if the sampling is at least twice the highest frequency on the channel, then the result sounds OK. (See NYQUIST THEOREM.) Thus, the highest frequency on a voice phone line is 4,000 Hertz. So one must sample it at 8,000 times a second. Many PCM digital voice conversations are typically put on one communications channel. In North America, the most typical channel is called the T-1 (also spelled T1). It places 24 voice conversations on two pairs of copper wires (one for receiving and one for transmitting). It contains 8000 frames each of 8 bits of 24 voice channels plus one framing (synchronizing bit) bit which equals 1.544 Mbps, i.e. 8000 x (8 x 24 + 1) equals 1.544 megabits.

Countries outside of the United States and North America use a different scheme for multiplexing voice conversations. It is based not on 24 voice channels, but on 32. This scheme keeps two of the 32 channels for control, actually transmitting 30 voice conversations at a data rate of 2.048 Mbps. The European system is calculated as 8 bits x 32 channels x 8000 frames per second. European PCM multiplexing is not compatible with North American multiplexing. The two systems cannot be directly connected. Some PBXs in the U.S. conform to the U.S. standard only. Some (very few) conform to both. Both the European and North American T-1 "standards" have now been accepted as ISDN "standards." In addition to PCM, there are many other ways of digitally encoding voice. PCM remains the most common. See T-1 and VOICE COMPRESSION.

PCX
PC Paintbrush file format.

PRESENTATION STATUS
For a particular call, an item that indicates if a calling identity item may be presented to the called party. If the presentation status is "public" presentation is allowed. If it is anonymous", presentation is restricted. Presentation status has to do with the presentation or not or calling line identification numbers.

PRI
Primary Rate Interface. The ISDN equivalent of a T-1 circuit. The Primary Rate Interface (that which is delivered to the customer's premises) provides 23B+D (in North America) or 30B+D (in Europe) running at 1.544 megabits per second and 2.048 megabits per second, respectively. There is another ISDN interface. It's called the Basic Rate Interface. It delivers 2B+D over either one

or two pairs. In ISDN, the "B" stands for Bearer, which is 64,000 bits per second, which can carry PCM-digitized voice or data. See ISDN for a much better explanation.

PSTN
Public Switched Telephone Network. An abbreviation used by the ITU-T, PSTN simply refers to the local phone company.

PTT
Post Telephone & Telegraph administration. The PTTs, usually controlled by their governments, provide telephone and telecommunications services in most foreign countries. In ITU-T documents, these are the Administrations referred to as Operating Administrations. The term Operating Administrations also refers to "Private Recognized Operating Agencies" which are the private companies that provide communications services in those very few countries that allow private ownership of telecommunications equipment.

PULSE DIALING
One or two types of dialing that uses rotary pulses to generate the telephone number. See ROTARY DIAL.

R1
The ITU-T's name for a particular North American digital trunk protocol that happens to use multi-frequency (MF) pulsing. Some Europeans refer to any North American

MF signaling protocol as R1 when distinguishing it from their own R2. See R2 and MF.

R2
A whole series of ITU-T specs which refers to European analog and digital trunk signaling. It refers to a type of trunk found in Europe which uses compelled handshaking on every MF (multi-frequency) signaling digit.

RBOC
Regional Bell Operating Company. There are seven RBOCs each of which own two or more BOCs (Bell Operating Companies). The RBOCs were carved out of the old AT&T/Bell System by Judge Harold Greene when he signed off on the divestiture of the Bell operating companies from AT&T at the end of 1983. There is nothing magical about seven -- nor the grouping of BOCs into RBOCs -- except the Judge wanted to keep them all roughly the same size. The seven RBOCs are Ameritech, Bell Atlantic, BellSouth, NYNEX, Pacific Telesis, Southwestern Bell and US West. In early October, 1994, Southwestern Bell changed its name to SBC Communications Inc. Its telephone companies, it said, would still operate under the Southwestern name.

RELATIONAL DATABASE
A database that is organized and accessed according to relationships between data items. A relational database consists of tables, rows and columns. In its simplest conception, a relational database is actually a collection of data files that "relate" to each other through at least one common field. For example, one's employee number can

be the common thread through several data files --
payroll, telephone directory, etc. One's employee number
might thus be a good way of relating all the files together
in one gigantic data base management system (DBMS).

RING
As in Tip and Ring. One of the two wires (the two are Tip
and Ring) needed to set up a telephone connection.

RING BACK TONE
The sound you hear when you're calling someone else's
phone. The tone you hear is generated by a device at your
central office and may bear no relationship to the sound
the phone at the other end is emitting -- or not emitting.

RING GENERATOR
A component of virtually all phone systems, ranging from
large central offices to small key systems, that supplies
the power to ring the bells inside phones, typically 90
volts AC at 20 Hz.

RING TRIP
The process of stopping the AC ringing signal at the
central office when the telephone being rung is answered.

RJ
Registered Jacks. They're telephone and data plugs
registered with the FCC. RJ-XX (where X is a number)
are probably the most common plugs in the world.

RJ-11

RJ-11 is a six conductor modular jack that is typically wired for four conductors (i.e. four wires). The RJ-11 jack (also called plug) is the most common telephone jack in the world. The RJ-11 is typically used for connecting telephone instruments, modems and fax machines to a female RJ-11 jack on the wall or in the floor. That jack in turn is connected to twisted wire coming in from "the network" -- which might be a PBX or the local telephone company central office. In a home installation, the red and green pair would be used for carrying the phone conversation and the black and white might be used for carrying low voltage from a plugged-in power transformer to light buttons on the phone. In many offices, the tip and ring were used for the voice conversation and the black and white were used for signaling. Increasingly, these days more and more office phone systems use only one pair, i.e. the red and green conductors.

RJ-14

A jack that looks and is exactly like the standard RJ-11 that you see on every single line telephone. Whereas the RJ-11 defines one line -- with the two center, red and green, conductors being tip and ring, the RJ-14 defines two phone lines. One of the lines is the "normal" RJ-11 line -- the red and green conductors in the center. The second line is the second set of conductors -- black and yellow -- on the outside.

RJ-21X

An Amphenol connector under a different name. Here's the explanation: Amphenol is a manufacturer of electrical and electronic connectors. They make many different

470

models, many of which are compatible with products made by other companies. Their most famous connector is probably the 25-pair connector used on 1A2 key telephones and for connecting cables to many electronic key systems and PBXs. The telephone companies call the 25-pair Amphenol connector used as a demarcation point the RJ-21X. The RJ-21X connector is made by other companies including 3M, AMP and TRW. People in the phone business often call non-amphenol 25-pair connectors, amphenol connectors.

RJ-48C
An 8-position keyed plug most commonly used for connecting T-1 circuits. The RJ-48C is an 8-position plug with four-wires (two for transmit, two for receive) commonly connected. When the phone company delivers T-1 to your offices, it usually terminates its T-1 circuit on a RJ-48C. And it expects you to connect that RJ-48C to your phone system or T-1 channel bank and then to your phone system.

ROBBED-BIT SIGNALING
Robbed bit signaling typically uses bits known as the A and B bits. These bits are sent by each side of a T-1 termination and are buried in the voice data of each voice channel in the T-1 circuit. Hence the term "robbed bit" as the bits are stolen from the voice data. Since the bits are stolen so infrequently, the voice quality is not compromised by much. But the available signaling combinations are limited to ringing, hang up, wink, and pulse digit dialing.

RS-232

RS for Recommended Standard. A set of standards specifying various electrical and mechanical characteristics for interfaces between computers, terminals and modems. The RS-232-C standard, which was developed by the EIA (Electrical Industries Association), defines the mechanical and electrical characteristics for connecting DTE and DCE data communications devices. It defines what the interface does, circuit functions and their corresponding connector pin assignments. The standard applies to both synchronous and asynchronous binary data transmission.

Most personal computers use the RS-232-C interface to attach modems. Some printers also use RS-232-C. You should be aware that despite the fact that RS-232-C is an EIA "standard," you cannot necessarily connect one RS-232-C equipped device to another one (like a printer to a computer) and expect them to work intelligently together. That's because different RS-232-C devices are often wired or pinned differently and may also use different wires for different functions. The "traditional" RS-232-C plug has 25 pins. The new IBM PC AT, most AT compatibles and the Toshiba T1100 Plus have a "new" RS-232-C plug with only nine pins. This smaller plug does essentially the same thing as its bigger cousin, but you need an adapter cable to connect one to another. They're widely available. See also DCE and DTE.

RS-328

October, 1966 the Electronic Industries Association issues its first fax standard: the EIA Standard RS-328, Message Facsimile Equipment for Operation on Switched Voice Facilities Using Data Communications Equipment. The

Group 1 standard, as it later became known, made possible the more generalized business use of fax. Transmission was analog and it took four to six minutes to send a page.

SAMPLING RATE
The number of times per second that an analog signal is measured and converted to a binary number -- the purpose being to convert the analog signal to a digital signal. The most common digital signal -- PCM -- samples voice 8,000 times a second.

SEIZURE
To access a circuit and use it, or make it busy so that others cannot use it.

SPEAKER DEPENDENT VOICE RECOGNITION
Technology capable of recognizing speech from a given user after completion of a vocabulary training and enrollment procedure. It is not voice verification although it is sometimes confused with this technology.

SPEAKER INDEPENDENT VOICE RECOGNITION
SIR or SIVR. Technology capable of recognizing any user without prior training or knowledge of the user. SIR is used to accept input from callers to voice processors where the callers are using rotary dial phones instead of touchtone phones. SIR can substitute for the numbers on the DTMF keypad and can add the benefit of a few basic voice commands, e.g., Yes, No, Help, etc.

Because computer processing demands are formidable with speaker independent recognition, accurate speaker independent products are created with limited vocabularies. In contrast, trainable or speaker dependent recognizers can feature larger vocabularies at lower prices. SIR has been slowly gaining acceptance in telephone applications. SIR is increasingly used in automated operator assistance applications. SIR will see increased use as system builders respond to pressures to provide voice processing functions to the enormous rotary phone installed base domestically and abroad.

SPEECH CONCATENATION
A term used in voice processing for economical digitized speech playback that uses independently recorded files of phrases or file segments linked together under application program control to produce a customized response in natural sounding language. For example, order status, bank balances, bus schedules or lottery results, etc. Concatenation is done for speed and economy. It lends itself to limited and structured vocabularies that are best stored in RAM (Random Access Memory) or speedily accessible from disk. Concatenation does not replace Text-To-Speech (TTS) as a method of getting the voice processor to deliver its responses.

SPEECH RECOGNITION
Voice recognition is the ability of a machine to recognize your particular voice. This contrasts with speech recognition, which is different. It is the ability of a machine to understand human speech -- yours and everyone else's. Voice recognition needs training. Speaker independent recognition does not require training.

474

STATION
A dumb word for a telephone. Also called an instrument, or a telephone instrument. An extension station is one connected "behind" a PBX or key system. In other words, the PBX or key system is between the station and the telephone central office. We tried to remove the word "station" from this dictionary, but failed. We suspect the word comes from the very old days when the telephone industry was regulated by the Interstate Commerce Commission, (the ICC) which also regulated the railroad industry.

SUBSCRIBER
A person or company who has telephone service provided by a phone company. In other industries, subscribers are called customers. Some telephone companies are beginning to call their subscribers customers. Thank goodness.

SUBSCRIBER LINE
The telephone line connecting the local telephone company's central office to the subscriber's telephone instrument or telephone system.

SWITCH
A mechanical, electrical or electronic device which opens or closes circuits, completes or breaks an electrical path, or selects paths or circuits.

SWITCH HOOK

It is also called the Hook Switch. A switch hook or hook switch was originally an electrical "switch" connected to the "hook" on which the handset (or receiver) was placed when the telephone was not in use. The switch hook is now the little plunger at the top of most telephones which is pushed down when the handset is resting in its cradle (on-hook). When the handset is raised, the plunger pops up (the phone goes off-hook). Momentarily depressing the switch hook (under 0.8 of a second) can signal various services such as calling the attendant, conferencing or transferring calls.

SYNCHRONOUS

The condition that occurs when two events happen in a specific time relationship with each other and both are under control of a master clock. Synchronous transmission means there is a constant time between successive bits, characters or events. The timing is achieved by the sharing of a single clock. Each end of the transmission synchronizes itself with the use of clocks and information sent along with the transmitted data. Synchronous is the most popular communications method to and from mainframes. In synchronous transmission, characters are spaced by time, not by start and stop bits. Because you don't have to add these bits, synchronous transmission of a message will take fewer bits (and therefore less time) than asynchronous transmission. But because precise clocks and careful timing are needed in synchronous transmission, it's usually more expensive to set up synchronous transmission. Most networks are synchronous these days. See ASYNCHRONOUS.

T-1

Also spelled T1. A digital transmission link with a capacity of 1.544 Mbps (1,544,000 bits per second). T-1 uses two pairs of normal twisted wires, the same as you'd find in your house. T-1 normally can handle 24 voice conversations, each one digitized at 64 Kbps. But, with more advanced digital voice encoding techniques, it can handle more voice channels. T-1 is a standard for digital transmission in the United States, Canada, Hong Kong and Japan.

T-1 lines are used for connecting networks across remote distances. Bridges and routers are used to connect LANs over T-1 networks. There are faster services available. T-1 links can often be connected directly to new PBXs and many new forms of short haul transmission, such as short haul microwave systems. It is not compatible with T-1 outside the United States and Canada. In Europe T-1 is called E-1 or E1.

Outside of the United States and Canada, the "T-1" line bit rate is usually 2,048,000 bits per second. France and West Germany impose slight variations that make their formats unique. Only one element remains constant -- the DS-0. The 64 kilobit per channel is universal. Most often it represents a PCM voice signal sampled at 8,000 times per second. However, the form of PCM encoding differs between T-1 (mu-law) and E-1 (A-law companding). According to Bill Flanagan's book, the differences are not so great that a multiplexer cannot convert between them. Conversion of E-1 to T-1 involves both the compression law and the signaling format.

T.30

ITU-T standard. Handshake protocol. This standard describes the overall procedure for establishing and managing communication between the two fax machines. There are five phases of operation covered: call set up, pre-message procedure (selecting the communication mode), message transmission (including phasing and synchronization), post message procedure (end-of-message and confirmation) and call release (disconnection). See CCITT.

TALK OFF

Talk off (also called Talk-Off) is one hazard of in-band signaling. The classic definition of talk-off is that it occurs when your voice has enough 2600 Hz energy to activate the 2600 Hz tone-detecting circuits in the central office. The 2600 Hz tone is used for in-band signaling. In voice processing, talk-off happens when a person is recording an audio file (say, leaving a voice mail message) and the frequencies in his voice happen to match those of a touch-tone digit. The voice board reacts as if a touchtone digit were press. Which means it may terminate the recording immediately and move to another menu selection.

TAPI

Telephone Application Programming Interface. Also called Microsoft/Intel Telephony API. A term that refers to the Windows Telephony API. TAPI is an evolving and improving set of functions supported by Windows that allow Windows applications (3.X, 95 and NT) to program telephone-line-based devices such as single and multi-line phones (both digital and analog), modems and fax cards in a device-independent manner. See also WOSA. TAPI essentially does for telephony devices what the Windows Print Manager did for printers.

TARIFF
Documents filed by a regulated telephone company with a state public utility commission or the Federal Communications Commission. The tariff, a public document, details services, equipment and pricing offered by the telephone company (a common carrier) to all potential customers. Being a "common carrier" means it (the phone company) must offer its services to everybody at the prices and at the conditions outlined in its public tariffs. Tariffs do not carry the weight of law behind them. If you or the telephone company violate them, no one will go to jail. The worst that can happen to you, as a subscriber, is that your service will be cut off, or threatened to be cut off. Regulatory authorities do not normally approve tariffs. They accept them -- until they are successfully challenged before a hearing of the regulatory body or in court (usually Federal Court). Many tariffs were accepted by regulatory commissions only to be struck down in court as unlawful, discriminatory, not cost-justified, etc. Monies collected under the tariff have been refunded and unnecessary equipment removed. In these new, competitive days, many telephone companies are violating their own tariffs by charging less money than their tariffs say they should, or bundling services together at a discount. They are also providing service and equipment on terms less onerous than outlined in their tariffs. Many users now regard tariffs as starting bargaining points, rather than ending bargaining points.

TCF

Training Check Frame. Last step in a series of signals in a fax transmission called a training sequence, designed to let the receiver adjust to telephone line conditions.

TDD

Telecommunications Device for the Deaf. Under the Communications Act of 1934, a TDD is defined as a machine "that employs graphic communication in the transmission of coded signals through a wire or radio." TDD devices (which typically look like simple computer terminals) use the Baudot method of communications. Most TDD devices are acoustically coupled and are slow, running at 300 baud. See DAUDOT CODE.

TDM

TIME DIVISION MULTIPLEXING. TDM. A technique for transmitting a number of separate data, voice and/or video signals simultaneously over one communications medium by quickly interleaving a piece of each signal one after another. Here's our problem. We have to transport the freight of five manufacturers from Chicago to New York. Each manufacturer's freight will fit into 20 rail boxcars. We have three basic solutions. First, build five separate railway lines from Chicago to New York. Second, rent five engines and schlepp five complete trains to New York on one railway track. Or, third, join all the boxcars together into one train of 100 boxcars and run them on one track. The train might look like this: Engine, Boxcar from Producer A, Box Car from Producer B, Producer C, Producer D, Producer E, and then the order begins again...Boxcar from Producer A, Producer B...Moving one large train of 100 boxcars is likely to be cheaper and more efficient than moving five smaller trains each of 20 boxcars on five separate railway tracks. Time Division Multiplexing, thus, represents substantial savings over have five separate networks (five separate tracks) and sending five separate transmissions (five separate trains).

This is what Time Division Multiplexing is all about. And the analogy is perfect. Take one large train (fast communications channel) and interleave pieces (boxcars) from each conversation one after another. If you do this fast enough, you'll never notice you've broken the conversations apart, moved them separately, and then put them back together at the distant end. In TDM, you "sample" each voice conversation, interleave the samples, send them on their way, then reconstruct the several conversations at the other end. There are several ways to do the sampling. You can sample eight bits (one byte) of each conversation, or you can sample one bit. The former is called word interleaving; the latter bit interleaving. The basic goal of multiplexing -- whether it be time division multiplexing, or any other form -- is to save money, to cram more conversations (voice, data, video or facsimile) onto fewer phone lines. To substitute electronics for copper.

TEXT-TO-SPEECH SYNTHESIS
TTS. Technologies for converting textual (ASCII) information into synthetic speech output. Used in voice processing applications requiring production of broad, unrelated and unpredictable vocabularies, e.g., products in a catalog, names and addresses, etc. This technology is appropriate when system design constraints prevent the more efficient use of speech concatenation alone. See SPEECH CONCATENATION.

481

TIFF
Tagged Image File Format. TIFF provides a way of storing and exchanging digital image data. Aldus Corp., Microsoft Corp., and major scanner vendors developed TIFF to help link scanned images with the popular desktop publishing applications. It is now used for many different types of software applications ranging from medical imagery to fax modem data transfers, CAD programs, and 3D graphic packages. The current TIFF specification supports three main types of image data: Black and white data, halftones or dithered data, and gray scale data. Some wags think TIFF stands for "Took It From a FotograF." It doesn't.

TIME OUT
In telecommunications and computer networks, an event which occurs at the end of a predetermined interval of time is called Time Out. For example if you lift the phone off the cradle and do not proceed to dial, after a certain number of seconds you will hear either a voice telling you to get on with it or a howling sound of some sort. Data networks have the same thing. Don't do anything for x minutes and the system will knock you off the air, i.e. hang up on you. In more technical terms, time out is the amount of time that hardware or software waits for an expected event before taking corrective action. In its most common form, time out is the amount of time an OCC or telephone system waits after your call goes through before it begins billing or timing the call.

TIME SLOT
1. In time division multiplexing (TDM) or switching, the slot belonging to a voice, data or video conversation. It

can be occupied with conversation or left blank. But the slot is always present. You can tell the capacity of the switch or the transmission channel by figuring how many slots are present.

2. An SCSA term. The smallest switchable data unit on the SCbus or Scxbus Data Bus. A time slot consists of eight consecutive bits of data. One time slot is equivalent to a data path with a bandwidth of 64 Kbps.

TIP
The first wire in a pair of phone wires. The second wire is called the "ring" wire. The tip is the conductor in a telephone cable pair which is usually connected to positive side of a battery at the telephone company's central office. It is the phone industry's equivalent of Ground in a normal electrical circuit. See TIP & RING.

TIP & RING
An old fashioned way of saying "plus" and "minus," or ground and positive in electrical circuits. Tip and Ring are telephony terms. They derive their names from the operator's cordboard plug. The tip wire was connected to the tip of the plug, and the ring wire was connected to the slip ring around the jack. A third conductor on some jacks was called the sleeve. That's it. Nothing more sinister. Nothing more interesting.

TONE DIAL
A pushbutton telephone dial that makes a different sound (in fact, a combination of two tones) for each number pushed. The correct name for tone dial is "Dual Tone

MultiFrequency" (DTMF). This is because each button generates two tones, one from a "high" group of frequencies -- 1209, 1136, 1477 and 1633 Hz -- and one from a "low" group of frequencies -- 697, 770, 852 and 841 Hz. The frequencies and the keyboard, or tone dial, layout have been internationally standardized, but the tolerances on individual frequencies vary between countries. This makes it more difficult to take a touchtone phone overseas than a rotary phone.

You can "dial" a number faster on a tone dial than on a rotary dial, but you make more mistakes on a tone dial and have to redial more often. Some people actually find rotary dials to be, on average, faster for them. The design of all tone dials is stupid. Deliberately so. They were deliberately designed to be the exact opposite (i.e. upside down) of the standard calculator pad, now incorporated into virtually all computer keyboards. The reason for the dumb phone design was to slow the user's dialing down to the speed Bell central offices of early touch tone vintage could take. Today, central offices can accept tone dialing at high speed. But sadly, no one in North America makes a phone with a sensible, calculator pad or computer keyboard dial. On some telephone/computer workstations you can dial using the calculator pad on the keyboard. This is a breakthrough. It a lot faster to use this pad. The keys are larger, more sensibly laid out and can actually be touch-typed (like touch-typing on a keyboard.) Nobody, but nobody can "touch-type" a conventional telephone tone pad. A tone dial on a telephone can provide access to various special services and features -- from ordering your groceries over the phone to inquiring into the prices of your (hopefully) rising stocks.

TOUCHTONE

Touchtone is not a trademark of AT&T, despite what editions one through six of Newton's Telecom Dictionary said. It is a generic term for pushbutton telephones and pushbutton telecommunications services and the term "touchtone" may be used by anyone. At one stage it was a trademark of AT&T. At divestiture in 1984, AT&T gave it to the public. For a full explanation of touchtone, see DTMF, which stands for Dual Tone Multi-Frequency signaling, i.e. touchtone. Also dee TONE DIAL.

TRANSDUCER

A device which converts one form of energy into another. The diaphragm in the telephone receiver and the carbon microphone in the transmitter are transducers. They change variations in sound pressure (your voice) to variations in electricity, and vice versa. Another transducer is the interface between a computer, which produces electron-based signals, and a fiber-optic transmission medium, which handles photon-based signals.

TRUNK

A communication line between two switching systems. The term switching systems typically includes equipment in a central office (the telephone company) and PBXs. A tie trunk connects PBXs. Central office trunks connect a PBX to the switching system at the central office.

TSAPI

Telephony Services Application Programming Interface. Described by Novell and AT&T (its originators) as a standards-based API for call control, call/device monitoring and querying, call routing, device/system maintenance capabilities and basic directory services.

TTS
See TEXT-TO-SPEECH SYNTHESIS.

V.17
New ITU-T standard for simplex (one-way transmission) modulation technique for use in extended Group 3 Facsimile applications only. Provides 7200, 9600, 12000, and 14400 bps trellis-coded modulation (the modulation scheme is similar to V.33), MMR (Modified Modified Read) compression and error-correction mode (ECM).

V.21
ITU-T standard for 300 bit per second duplex modems for use on the switched telephone network. V.21 modulation is used in a half-duplex mode for Group 3 fax negotiation and control procedures (ITU-T T.30). Modems made in the U.S. or Canada follow the Bell 103 standard. However, the modem can be set to answer V.21 calls from overseas.

V.27
ITU-T standard for 4,800 bits per second modem with manual equalizer for use on leased telephone-type circuits. May be full-duplex on four wire leased lines, or half-duplex on two wire lines.

V.29
ITU-T standard for 9,600 bits per second modem for use on point-to-point leased circuits. Virtually all 9,600 bps leased line modems adhere to this standard. V.29 uses a carrier frequency of 1700 Hz which is varied in both phase and amplitude. V.29 also provides fallback rates of 4800 and 7200 bps. V.29 can be full-duplex on 4-wire leased circuits, or half-duplex on two wire and dial up circuits. V.29 is the modulation technique used in Group 3 fax for image transfer at 7200 and 9600bps.

VAR
Value Added Reseller. Typically VARs are organizations that package standard products with software solutions for a specific industry. VARs include business partners ranging in size from providers of specialty turn-key solutions to larger system integrators.

VOICE COMPRESSION
Refers to the process of electronically modifying a 64 Kbps PCM voice channel to obtain a channel of 32 Kbps or less for the purpose of increased efficiency in transmission.

VRU
Voice Response Unit. Think of a Voice Response Unit (also called Interactive Voice Response Unit) as a voice computer. Where a computer has a keyboard for entering information, an IVR uses remote touchtone telephones. Where a computer has a screen for showing the results, an IVR uses a digitized synthesized voice to "read" the screen to the distant caller. An IVR can do whatever a computer can, from looking up train timetables to moving

calls around an automatic call distributor (ACD). The only limitation on an IVR is that you can't present as many alternatives on a phone as you can on a screen. The caller's brain simply won't remember more than a few. With IVR, you have to present the menus in smaller chunks.

WINK

A signal sent between two telecommunications devices as part of a hand-shaking protocol. It is a momentary interruption in SF (Single Frequency) tone, indicating that the distant central office is ready to receive the digits that have just been dialed. In telephone switching systems, a single supervisory pulse. On a digital connection such as a T-1 circuit, a wink is signaled by a brief change in the A and B signaling bits. On an analog line, a wink is signaled by a change in polarity (electrical + and -) on the line.

WINK START

Also called Wink Operation. A timed off-hook signal normally of 140 milliseconds, which indicates the availability of an incoming register for receiving digital information from the calling office. A control system for phone systems using address signaling.

WOSA

Windows Open Services Architecture. According to Microsoft, WOSA provides a single system level interface for connecting front-end applications with back-end services. Windows Telephony, announced in May 1993, is part of WOSA. According to Microsoft, application developers and users needn't worry about conversing with numerous services, each with its own protocols and interfaces, because making these connections is the business of the operating system, not of individual applications. WOSA provides an extensible framework in which Windows based applications can seamlessly access information and network resources in a distributed computing environment. WOSA accomplishes this feat by making a common set of APIs available to all applications. WOSA's idea is to act like two diplomats speaking through an interpreter. A front-end application and back-end service needn't speak each other's languages to communicate as long as they both know how to talk to the WOSA interface (e.g. Windows Telephony). As a result, WOSA allows application developers, MIS managers, and vendors of back-end services to mix and match applications and services to build enterprise solutions that shield programmers and users from the underlying complexity of the system.

This is how WOSA works: WOSA defines an abstraction layer to heterogeneous computing resources through the WOSA set of APIs. Initially, this set of APIs will include support for services such as database access, messaging (MAPI), file sharing, and printing. Because this set of APIs is extensible, new services and their corresponding APIs can be added as needed.

WOSA uses a Windows dynamic-link library (DLL) that allows software components to be linked at run time. In this way, applications are able to connect to services dynamically. An application needs to know only the definition of the interface, not its implementation. WOSA defines a system level DLL to provide common procedures that service providers would otherwise have to implement. In addition, the system DLL can support functions that operate across multiple service implementations. Applications call system APIs to access services that have been standardized in the system. The code that supports the system APIs routes those calls to the appropriate service provider and provides procedures and functions that are used in common by all providers.

The primary benefit of WOSA is its ability to provide users of Windows with relatively seamlessly connections to enterprise computing environments. Other WOSA benefits, according to Microsoft include:

- Easy upgrade paths.
- Protection of software investment.
- More cost-effective software solutions.
- Flexible integration of multiple-vendor components.
- Short development cycle for solutions.
- Extensibility to include future services and implementations.

ZEROFILL
A traditional fax device is mechanical. It must reset its printer and advance the pages as it prints each scan line it receives. If the receiving machine's printing capability is slower than the transmitting machine's data sending capability, the transmitting machine adds "fill bits" (also called Zero Fill) to pad out the span of send time, giving the slower machine the additional time it needs to reset prior to receiving the next scan line.

Appendix A|
Vendor Contact Information

Software Vendors

Expert Systems, Inc.
1301 Hightower Trail
Suite 201
Atlanta, GA 30350
Tel: 770.642.7575
Fax: 770.587.5547
Email: info@EASEy.com

Microsoft Corporation
Microsoft Developer Network
One Microsoft Way
Redmond, WA 98052-6399
Tel: 800.759.5474
Fax: 206.936.7329, Attn: Developer Network

Voice Information Systems, Inc.
2118 Wilshire Blvd.
Suite 973
Santa Monica, CA 90403
Tel: 800.234.VISI
Fax: 800.234.FXIT

Hardware and CT Component Vendors

Aculab, Ltd.
Aculab House
Old Road, Leighton Buzzard
Beds, LU7 7RG   England
Tel: +441.525.371393
Fax: +441.525.381284

Aerotel Ltd.
5 Hazoref Street
Holon 58856, Israel,
Tel. +972.3.559.3222
Fax +972.3.559.6111.

Berkeley Speech Technologies, Inc.
2246 Sixth Street
Berkeley, CA 94710
Tel: 510.841.5083
Fax: 510.841.5093

Dialogic Corporation
1515 Route 10
Parsippany, NJ 07054-4596
Tel: 800.755.4444
Fax: 201.993.3093

Dianatel Corporation
96 Bonaventure Drive
San Jose, CA 95134
Tel: 408.428.1000
Fax: 408.433.3388

GammaLink
1314 Chesapeake Terrace
Sunnyvale, CA 94089
Tel: 408.744.1400
Fax: 408.745.2375

Interface Systems, Inc.
5855 Interface Drive
Ann Arbor, MI 48103
Tel: 313.769.5900
Fax: 313.769.1047

Micro-Integration Corp.
One Science Park
Frostberg, MD 21532
Tel: 800.600.6448
Fax: 301.689.0808

Voice Control Systems, Inc.
14140 Midway Road
Dallas, TX 75244
Tel: 214.386.0300
Fax: 214.386.5555

Consulting Services

The Kauffman Group
324 Windsor Drive
Cherry Hill, NJ 08002
Tel: 609.482.8288
Fax: 609.482.8940

Index

B

C

E

F

G

H

N

O

P

R

S

Special Book Offer

Fax to: +1 770.587.5547

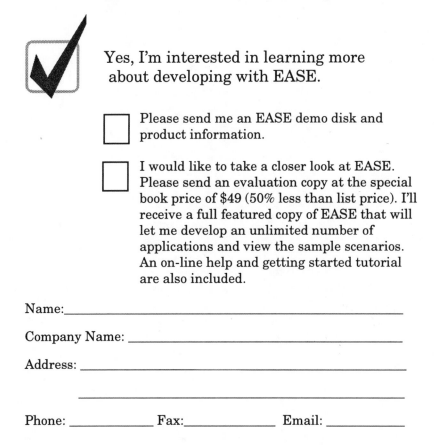

Yes, I'm interested in learning more about developing with EASE.

☐ Please send me an EASE demo disk and product information.

☐ I would like to take a closer look at EASE. Please send an evaluation copy at the special book price of $49 (50% less than list price). I'll receive a full featured copy of EASE that will let me develop an unlimited number of applications and view the sample scenarios. An on-line help and getting started tutorial are also included.

Name:_____

Company Name: _____

Address: _____

_____

Phone: _____ Fax:_____ Email: _____

Expert Systems, Inc.
1301 Hightower Trail • Suite 201 • Atlanta, GA 30350 USA
Tel: +1 770.642.7575 • Fax: +1 770.587.5547 • Email: info@EASEy.com